Green Energy and Technology

For further volumes:
http://www.springer.com/series/8059

Maria Kopsakangas-Savolainen
Rauli Svento

Modern Energy Markets

Real-Time Pricing, Renewable Resources and Efficient Distribution

Springer

Maria Kopsakangas-Savolainen
Finnish Environment Institute
and Thule Institute
University of Oulu
Asematie 10
90850 Martinniemi
Finland

Rauli Svento
Department of Economics
Martti Ahtisaari Institute of Global
Business and Economics
University of Oulu
Annantie 9 C 8
90560 Oulu
Finland

ISSN 1865-3529 e-ISSN 1865-3537
ISBN 978-1-4471-2971-4 e-ISBN 978-1-4471-2972-1
DOI 10.1007/978-1-4471-2972-1
Springer London Heidelberg New York Dordrecht

British Library Cataloguing in Publication Data
A catalogue record for this book is available from the British Library

Library of Congress Control Number: 2012933842

Printed on acid-free paper

Springer is part of Springer Science+Business Media (www.springer.com)

To Harri, Volmari, Vilhelmiina, Magdaleena and Pietari

Maria Kopsakangas-Savolainen

To Maria-Liisa and Elisa

Rauli Svento

Preface

Energy has moved to the forefront in societal and economic development. Wise economic decisions are needed for the questions of the use of non-renewable and renewable energy sources, pollution and global warming. Our homes and real estate are all the time more and more dependent on electricity. Electricity bills take a growing share of the budgets of households and firms, and this development creates new needs for smart usage of electricity.

Electricity industries have been among the first for deregulation and liberalization and room has been given for market mechanisms. We are currently in the situation to critically evaluate this development. Technology constraints for the use of efficient market mechanism are vanishing and this enables the use of new models such as incentive oriented real-time pricing.

While generation has been deregulated transmission and distribution are still, because of their natural monopoly, being regulated. New economic theory based, incentive driven regulation mechanisms have, however, been developed and practically applicable versions already exist. We are currently beginning to see significant structural changes also in power network systems. This change relates to intelligent networks or smart grids as they are also called. The basic element that relates to intelligent networks is that they change the one way traffic still going on in dumb networks to a two way dynamic system. The future smart grids allow the role of consumers to change from passive out takers to active users and optimizers of their extended energy possibilities. This creates new challenges and possibilities for the whole chain of the power system.

There is a strong and growing need to understand the energy market in a comprehensive manner. This means that all parts of the whole power system chain must be analyzed at the same time. The flow of electricity from generation through transmission and distribution to the final consumer, and the roles of all the players in this market, form an interesting entirety for economic analysis. The main motivation of this book is to give a comprehensive economically oriented picture to this extremely interesting and central field of modern societies.

We hope that this book is good reading for economics and engineering students as well as researchers interested in environmental and energy issues. The book also covers timely and relevant issues related to societal and economic decision making and thus is good reading also for officials and decision makers in environment and energy related fields.

We thank our colleagues at the Department of Economics at the University of Oulu, at the Finnish Environment Institute, at the Thule Institute and at the Martti Ahtisaari Institute of Global Business and Economics.

We thank Yrjö Jahnsson Foundation, Fortum Foundation and Martti Ahtisaari Institute of Global Business and Economics for financial support for our research.

Anthony Doyle and Grace Quinn of Springer provided valuable comments and editorial help.

Our deepest thanks go to our families. They gave us support and encouragement to start and finalise the writing process of this book.

Contents

Chapter 1
Introduction

Energy and electricity have become key elements and drivers in the modern world. In our networked economies with digital products and processes and device-driven consumption stabile, riskless and justly priced electricity supply is the current day basement upon which everything else more or less builds on. At the same time the use of many primary energy sources can be connected with greenhouse gas emissions and global warming. This basement has also gone through heavy structural changes and even bigger changes can be seen in the future. In this book we try to envision and analyze these changes with an economist's eye.

The original way of functioning of the power system was rather straightforward. This functioning was based on one-way transport of electricity from large generators through transmission and distribution to individual consumers. The first wave of structural changes started in the latter part of the last century and was based on the research results showing that the basic belief of economies of scale for quite large volumes in generation did not materialize any more. This finding led to two major changes. These were deregulation of generation and retail and unbundling of generation, transmission, and distribution.

The basic new element that deregulation and liberalization brought into the system was that the physical delivery of electricity could be differentiated from its buying and selling. Consumers are able to make contracts with generators far from their regional environments. The electricity they take out of the power system based on these contracts is of course not the same physical entity that the contracted generator puts in the system but the system balancing mechanisms take care of the needed equilibrium between supply and demand as a whole. The efficiency targets that can be expected from this development are based on two ingredients: competition in generation and retail and price sensitivity in consumption.

Competition in generation is increased because generators are able to make offers for contracts to all consumers. A lot of research has concentrated on the

M. Kopsakangas-Savolainen and R. Svento, *Modern Energy Markets*,
Green Energy and Technology, DOI: 10.1007/978-1-4471-2972-1_1,

question whether there exist enough generators even at the national level so that they cannot exercise market power in pricing. The answer to this problem has been either to demerger the big players or to form international pools for executing the merchandise.

Opening the generation field to competition combined with the question of greenhouse gases and global warming has of course brought the question of diversification onto the table. Renewable energy sources are currently taking their market shares as technology and market conditions emerge. But can renewable energy sources fare in the competition without being subsidized as is currently happening? Do they have a sustainable economic future also in the long run?

The price sensitivity of consumers of electricity has generally been rather low. This is understandable since the possibilities for adjustments in demand have not been elaborated and the price contracts have traditionally been based on flat rates so that the volatility of the real-time price has no direct effect on the bill. In generation, however, the cost of generation varies greatly depending on the amount of demand. The merit order of the power system is based on the idea of using most cost-efficient technologies first. While the most expensive peak technologies during the demand peaks are used, the prices are much higher than during the low demand periods. These generation cost-based differences do not reach the consumers directly.

Economists have for long seen this unbundling of marginal costs from current prices as potentially problematic. Technological constraints have until recent times prevented the use of the efficient price setting logic. The development and installation of intelligent automated metering technologies and devices, however, changes this picture. The first real-time pricing (RTP) contracts have already been signed and this development can be expected to proceed with space in the near future. But what kinds of effects will RTP have on generation and consumption and consequently on emissions? How is the generation system going to change? Can we get rid of expensive peak demand capacities? Is the generation system turning more efficient? How big changes in demand can be expected? How do consumer bills look after pricing is based on real-time prices?

Unbundling of generation, transmission, and distribution relates to the idea that even though generation can be exposed to competition the use of networks still has features of natural monopoly related to it. Being natural monopolies the distribution utilities can exercise market power and by unbundling the avoidance of cross-subsidization is pursued. But the avoidance of possible cross-subsidization is not the only problem related to natural monopolies. The potential for market power still exists and this necessitates regulation to be exercised over these utilities.

There are two things that need to be solved when designing efficient regulation mechanisms for distribution utilities or any other natural monopolies as well. The first relates to measurement of the efficiency of the agents in question. Since the regulation mechanism usually relates to pricing possibilities there must be a clear and evidence-based link between the true costs of the agent and the price it is allowed to set. Efficiency measurement has proceeded fast in recent years and current research is able to take many kinds of heterogeneities among the studied population into consideration. This is especially necessary concerning distribution utilities since they often operate in very different environments.

The other key question related to regulation is the need to make it modern in the sense of allowing it to be strongly incentive based. This changes the regulation setup to a game playing field where all actors must be active and work their optimal strategies thoroughly out. At the same time it is vital that the data that is being used as a basis for the game is objective in the above-mentioned heterogeneity accounting sense. What do the heterogeneity allowing incentive-based regulation models look like?

We are currently beginning to see the first signs of the second wave of structural changes in power systems. This change is once again related to technological possibilities that have been created and are becoming commercially applicable. Here, we refer to intelligent networks or smart grids as they are also called. The basic element that relates to intelligent networks is that they change the one-way traffic still going on in dumb networks to a two-way dynamic system. The future smart grids allow the role of consumers to change from passive out takers to active users and optimizers of their extended energy possibilities. Intelligent homes and other real estate can generate electricity over own needs and intelligent networks make it possible to feed this excess supply into the network. The energy environment changes into a distributed generation endowment where each player can have several roles. This naturally changes the whole business logic of local generation and once these changes gain decent volumes even the aggregate power system becomes viable. This new business logic necessitates a completely new role for new kinds of services related to installing, educating, and using all these new possibilities in intelligent ways. At the same time, of course, new possibilities for renewable energy sources open up.

These are the core questions that the modern energy markets are facing. We try to give our answers to them in this book and we hope that this book gives a comprehensive, analytical, and thought-provoking picture of the vast changes and possibilities that the world is facing in one of our basic fundamentals of modern life, i.e., electricity.

This book is organized so that after providing a general discussion on the deregulation and liberalization of the electricity industry we proceed step-by-step, analyzing the whole energy market system. We start with electricity production which is followed by analyses of distribution networks and regulation and conclude with the characterization of future electricity markets.

The more detailed structure of the book is as follows: In Chap. 2 we present and discuss the grounds and incentives of the deregulation and liberalization processes that have been carried out in many countries during the past few decades. We also assess the crucial factors which affect the potential successfulness of the deregulation and liberalization processes. The bibliographies are not intended to provide a complete survey of the relevant literature but rather to summarize the core research as a background for the subject of this book.

Chapters 3–6, concentrate on electricity production. In Chap. 3 we derive one of the various ways to model the energy production system. The main motivation to choose the approach used in this chapter is our goal of analyzing the energy market in a forward-looking sense. Especially, we want to include into the model the main characteristics of the future demand-side management. One way to model demand side is to include into the model the RTP mechanism. In Chap. 4 we analyze the potential effects of RTP of electricity on; the need for total, peak, and mid-merit capacities; total demand; prices; peak demand hours; and economic welfare in the Nordic power markets. The energy production market model is enlarged in Chaps. 5 and 6; first to analyze the potential impact of increase in nuclear power capacity on market factors and then the impact of different carbon emission prices as the promotion mechanism for renewable energy (wind) to enter on the market.

In Chaps. 7 and 8, we concentrate on the electricity distribution network. Electricity distribution is a natural monopoly industry and consequently there is a need for regulation. Efficiency measurement of distribution utilities is essential to achieve accurate information about the ingredients for efficient regulation. Because, for example, due to utility heterogeneities, efficiency measurement is a demanding task, here we analyze the various ways to perform cost efficiency analyses. Chapter 9 concentrates on the necessary regulation related to electricity networks. We first present and discuss the different theoretical regulation models and then we empirically compare the welfare effects of different regulation schemes of electricity distribution utilities.

The challenging developments of the new energy system that have been analyzed in this book necessitate a change from the traditional “dumb grid” to an intelligent and adaptive “smart grid”. This change in both transmission and distribution grids is well under way in many countries and large-scale effects of this transformation can be expected in the near future. This book concludes with the summarization and discussion of the basic economic features of this change.

Chapter 2
Restructuring of Electricity Markets

2.1 Liberalization, Deregulation, and Restructuring of the Electricity Markets

Governments have regarded the electricity industry as a leading industrial sector throughout history. Because of its strategic importance for industrial development, its impacts on social and environmental issues, and its natural monopoly characteristics, it has been seen necessary to regulate electricity industry effectively. Many countries have relied on public ownership of electricity supply assets instead of strict direct regulation. On the other hand, in countries with substantial private ownership since the early electrification, governments have typically subjected electric utilities to wide-ranging financial, health and safety, planning, and environmental control. These two approaches to the industry, public and private control, have ensued large-scale investments in costly technologies,[1] concentration on engineering excellence instead of cost minimization and high quality service, and lack of competition in the potentially competitive generation and supply businesses. Exceptions to these general rules can be found—for example the Scandinavian small-scale electricity distribution by municipally owned utilities.

Historically, the electricity industry has been characterized by economies of scale in the generation and necessity of an extensive transmission and distribution network in order to deliver the generated electricity to the final consumers. These primary components of electricity supply were integrated within individual electric utilities. However, in the mid-1980s it was realized in several countries that even though transmission and distribution networks are natural monopolies, the scale economies in electricity production at the generating unit level had exhausted at a unit size of about 500 MW (see e.g., [21, 24, 41]). This meant that the natural monopolistic characteristics of electricity supply and generation had vanished and thus they had become potentially competitive activities. As a consequence, it was

[1] Like, for example, nuclear power.

M. Kopsakangas-Savolainen and R. Svento, *Modern Energy Markets*, Green Energy and Technology, DOI: 10.1007/978-1-4471-2972-1_2,

noted that a separation of network activities from generation and supply and the introduction of competition to the potentially competitive parts of the industry might increase the overall efficiency.

It is possible to organize the competitive wholesale trading by using many different systems from which the pool-based trade and bilateral trade have become the most common. However, a certain degree of central co-ordination is needed because competitive wholesale trading arrangements all share the same need to match supply and demand, and this matching process must be carried out instantaneously.[2] This is true regardless of whether the electricity industry consists of a single vertically integrated public sector utility or a multitude of competing generators and suppliers.

The objective of this chapter is to discuss the grounds and incentives of deregulation and restructuring processes in the electricity industry. Further, the success of already implemented deregulation processes is assessed by using the Nordic power market as an example. Also the crucial factors in improving efficiency are determined.

2.1.1 On Liberalization and Deregulation

During the past two decades we have seen comprehensive electricity sector liberalization and deregulation in all EU countries. The same is not true for the U.S. since it has not enacted mandatory federal restructuring and competition law. In the U.S. any significant restructuring reforms have been left under the decision of the individual states. In consequence, many states have introduced only some liberalization reforms concerning mostly the wholesale markets. Actually, some of those states that have introduced more comprehensive restructuring and reforms on the electricity sector are now planning to re-regulate the industry (see [24]).

When evaluating the degree of reforms in different countries it should be noted that the concept of liberalization or restructuring may take several different forms. It may mean permission for independent generators to enter the market, the creation of a power pool, or the horizontal separation of incumbent generators. In addition, it can refer to the vertical disintegration of state-owned monopolies into generation, transmission, and distribution businesses. In its most comprehensive form, liberalization usually culminates in the sale of the state-owned assets, either completely or at least partially, to the private sector (see [39]). Joskow [24] gives a comprehensive list of the desirable features for restructuring and regulatory reform.

It is often argued that liberalization, and as an endpoint of it, privatization, improves the economic efficiency. The reason why liberalization and privatization are assumed to improve economic efficiency and how significant improvements they create are explained in different ways, depending on the theoretical basis

[2] This is because power cannot be stored economically in significant quantities.

adopted. The property rights theory (following [1], see also [39]) argues that privatization assigns particular assets to those who can utilize these assets most efficiently. The supporters of this theory claim that state-owned electricity utilities are not run as efficiently as they could be run under private ownership. This is basically due to the fact that the state-owned firm is not supposed to minimize costs as would be the case as a result of privatization. Bureaucracy theories (see [36]) argue that managers in state-owned companies may be more interested in maximizing the budget of their department than in minimizing costs or maximizing profits. On the other hand, the theory of regulation and incentives does not support privatization as strongly as the two theories above. As a matter of fact, the famous Averch and Johnson study [4] argues that in industries where privatized activities are regulated, the regulation may introduce negative incentives, which may not be present in the public sector and which would reduce economic efficiency.[3] More modern theories of regulation (see, e.g., [5, 30–32]) stress the importance of the information problems connected to the regulators' imperfect information about the true costs of the firm. The theory of influence activities asserts that ownership arrangements evidently change the relationships between groups and also their possibilities to influence within the company. These changes create some costs, which should be emphasized when planning privatization. Joskow [24] recognizes also many significant potential benefits, but also potential costs connected to the liberalization if the reforms are implemented incompletely or incorrectly. Green [15] emphasizes the importance of market power mitigation in order to reach significant efficiency gains as a result of liberalization. Concerning privatization the final effect can be positive since influence seeking is seen to be easier in private companies. However, as Newbery and Green [34] argue the relative performance of the industry does not depend strictly on whether the industry is under public or private ownership,[4] but rather on the state of the development of the industry, on technology, and on the balance of political and economic forces shaping its development.

Although privatization may have a positive influence on the performance of a firm, it should be emphasized that it can also create some problems. The potential problems arising from privatization include the high cost of regulation,[5] and the possibility of deadweight losses[6] if the privatized company can exploit market power. Because of these contradictory conceptions of privatization, Pollitt [38]

---

[3] Averch and Johnson [4] analyze the effect of rate-of-return regulation in the USA. They show how it creates incentives to over-invest in relation to the social optimum. They also argue that the rate-of-return regulation provides no incentive to reduce costs.

[4] Public ownership may be preferable when we deal with issues, such as coordination and restructuring, while private ownership may have comparative advantages considering the competition and self-centered objectives of the firm.

[5] The costs of regulation may include direct costs and also the costs resulting from poorer incentives for efficient performance.

[6] Deadweight losses may be due to the high prices, the social waste of entry-deterring activities, or the excessive entry caused by the high profits of incumbents (see, e.g., [39]).

argues that it is more an empirical issue rather than a theoretical one, whether the privatization process ultimately means lower costs and improved efficiency.

It should be pointed out that deregulation and liberalization are not simply a matter of public versus private ownership. Liberalization of markets has been done in different ways in different countries. Some countries have deregulated the industry by introducing competition and "stopping" the regulation while others have at the same time privatized the industry. Thus, we can say that the debate on public versus private ownership is more like a matter of choosing the modes of control. Liberalization itself includes subjecting utilities to market forces, which can result in more changes in performance than privatization. Replacing the monopoly activities by competition can increase efficiency. However, it should be noted that at the same time liberalization also redistributes rents and raises new regulatory problems in managing the interface between the regulated and the competitive parts of the utility (see, e.g., [33]). As Joskow [26] argues market imperfections (and the costs it causes) should be always evaluated with the regulatory imperfections (and the costs which it causes) when deregulating, regulating, or restructuring the market.

As already stated above, a central issue in creating the new electricity industry structure has been the observation that even though regulation or public ownership is the only stable form of organization in natural monopolies, potentially competitive parts can be separated from network parts. However, before restructuring can be thought to be complete there is the crucial question of how to combine the necessary regulation of the network with the organization of competition in activities that use the network as an input and are potentially competitive (see [32]). The issue of practical implementation of efficient regulation is still an open question (see, e.g., [29] and Chap. 8 in this book). Although an increasing number of countries have moved toward a more incentive-based price regulation, in most countries the basis for regulation is still based on the cost of supplying electricity, including an appropriate level of return on capital investments. The problems continually faced by regulators are how to determine "proper" costs, what is the appropriate depreciation rate of capital, and whether it is permissible to allocate more costs to one group of customers than to another (see e.g., [22, 23]). A further issue that has been seen as a threat to the success of deregulation is the possibility of some companies to exploit market power.

As one can conclude from the discussion above the electricity market deregulation, liberalization, and restructuring are not easy tasks. In consequence, even though we have seen successes in the electricity sector restructuring in countries like UK, the Nordic countries, Argentina, Chile, Texas, and portions of Australia, in many countries electricity sector reforms are moving forward slowly with considerable resistance or in some cases even moving backward [24].

2.1.2 Different Grounds for Deregulation

There are at least two fundamental reasons acting as the impetus for deregulation. First, deregulation can be based on changes in the ideological atmosphere.

This kind of a basis for deregulation usually culminates in the privatization of public activities. This has been argued to be the driving force for the deregulation process, for example, in the United Kingdom, where during the term of Margaret Thatcher, many industries, including the electricity industry, commenced restructuring. The number of producers has not been seen politically as critical as the privatization in order to reach the target of efficiency improvements as a result of deregulation. However, economists in UK argued in the early stages of the restructuring process that the number of generators in electricity markets should be higher than it was in England and Wales (see [17, 19, 40]).

Another ground for deregulation is based on the pure target to improve efficiency. In restructuring processes based on the pure efficiency target the number of operators (buyers as well as sellers) in the market has been seen as a crucial element, and not the privatization, in order to reach the target. The Nordic Electricity market is an example of this kind of a restructuring process. For example, there were nearly 340 market participants in the Nord Pool Spot in 2010. Although privatization has taken place, significant amount of the generators have remained in public ownership even after deregulation. The crucial element, in addition to the number of operators, has been seen to be the actuality of the demand function used by the pool operator. In the Nordic Power Market the demand function is calculated on the basis of the real bids to the pool instead of estimation by the pool operator. The more efficient allocation of production capacity has also been one of the motives in deregulating the Nordic Power markets.

Generally speaking, even though the political forces behind the decision to change the market conditions have been strong and varied in many countries, it has been argued (see, e.g., [11]) that deregulation would not have occurred if economists had not supported it through their research. Recently, economists have developed a theoretical and empirical framework to predict the actual effects of deregulation and liberalization. Just to mention a few studies, Wolak [42] has pointed out through international experiences the importance of efficient market monitoring in order to reach the benefit of deregulation and Green [15, 16] has studied the main characteristics and potential problems of competition policy in the European electricity market.

Although the potential benefits from deregulation are well known (see [8, 24]), there is no worldwide agreement upon the set of market rules for guaranteeing a successful industry restructuring. However, economists generally agree that because technological changes have frequently lessened the presence of scale economies, the prevalence and importance of natural monopoly features of the industry are diminishing. Already in [6] Baumol et al. argued in their theory of contestable markets that deregulation may be superior to regulation even in industries with scale economies. The contradictory opinions are related to the questions of how the deregulation should be implemented, and which kind of market rules should be created. It is clear that in some industries, such as electricity distribution and transmission, characteristics of natural monopoly and scale economies are so evident that most of the countries still rely on some form of regulation. There are varieties of methods to regulate the firms from which the so-called high-powered incentive regulation schemes are becoming more and more important. There is a lot of empirical evidence that the high-powered incentives created by competitive wholesale electricity networks will or have led to lower generator operating costs and also improved availability (see [10, 13, 23, 35]).

Although the major rationale for electricity industry restructuring is to provide stronger incentives for efficient production and delivery of electricity, it may not mean lower electricity prices if the firms possess market power and thus have the ability to raise output prices above the competitive levels. Consequently, it has to be decided which one of the two regimes will yield greater benefits to the final consumers: (1) a competitive market with strong incentives for least-cost production but limited incentives for cost-reflective output prices, or (2) a regulated market with limited incentives for least-cost production but potentially more cost-reflective output prices (see [8]).

The prevailing view is that the technologies for electricity generation and retailing are both such that competition is feasible. As discussed above, economies of scale in generation are exhausted at levels of production significantly below the current levels of industry output. However, the problem is how to guarantee that the price for electricity is set from the perspective of economic efficiency, i.e., such that it is set to mimic the market price in a competitive industry with many non-colluding firms and small barriers to entry.

2.2 Nordic Power Market as an Example of Restructuring

The Nordic power market, including Denmark, Finland, Norway, Sweden, and Estonia, provides a good example of restructuring and deregulating the electricity industry since 74% of all power in the region was traded in Nord Pool Spot in 2010. This makes the Nordic power market the world's largest market for buying and selling electric power.

The historical background of the electricity industry is fairly similar in all Scandinavian countries. Throughout the history of the industry there has been both public and private ownership of electricity companies. Another characteristic has been the relatively weak formal government-enforced regulation. Instead, there has been self-enforced club-regulation and yardstick competition. Also, the role of a publicly owned dominant firm has been extensive.[7] In addition, the share of hydropower has been and is relatively large in all Scandinavian countries except for Denmark.

The first commercial, relatively large-scale private power companies were established in the late nineteenth century. After that many local co-operatives were built, but the real expansion of the retail distribution of electricity took place

[7] At least in Norway, Sweden, and Finland.

shortly after the First World War. The next distinct stage of development in the electricity system was put forward after the Second World War. During the war, it had become clear that import and export of fuels were extremely difficult, which gave an incentive to further develop domestic hydropower in Scandinavia. As a result, hydropower capacity was increased rapidly in the 1940s and the 1950s. In the 1960s the expansion of hydropower slowed down, because the potential for unexploited hydro capacity was reduced. The increasing interest in environmental issues also changed the focus of future production and capacity exploitation from hydro to other alternatives (see [3, 18, 20]).

In Denmark, Finland, and Sweden several municipalities developed district-heating cogeneration systems based on oil, coal, biomass, or peat in the 1960s. Especially, Finland also built industrial cogeneration plants. The baseload production of electricity leaned on hydropower in Norway, coal, oil, and nuclear power in Finland, coal and oil in Denmark, and hydro and nuclear power in Sweden. The proportion of nuclear power was clearly increased in Sweden and Finland between the 1960s and the mid-1980s. In 1963, a co-operation organization, Nordel, was established. It enabled the collaboration between large generators in Denmark, Finland, Iceland, Norway, and Sweden. In practice, co-operation has been possible through high voltage direct current (HVDC) cables, which were constructed from Jutland to Sweden and Norway (see [18, 20]).

The retail distribution in urban areas was, already at the early stage of electrification, handled by utilities owned by towns or cities. In rural areas distribution co-operatives took the responsibility for retail distribution. As a result, there were numerous small and inefficient distributors in the mid-1940s. This problem was solved by regulation and nationalization, which resulted in a significant decline in the number of distributors, for example in Sweden from 2000 (in the mid-1950s) to 300 in 1996 (see [20]).

A common characteristic to Norway, Sweden, and Finland is that the population is concentrated in the south while the most of the hydro resources are in the north. As a result, transmission networks have been seen as very important since the first decades of the twentieth century.

2.2.1 Restructuring and Integration of the Nordic Power Markets

The deregulation of the Nordic electricity markets started in Norway on January 1991, as a new Energy Act was made effective. Originally, the Nord Pool was a national Norwegian power exchange, but it was expanded to cover also Sweden in 1996. It was extended further in 1998 when Finland joined the pool. In Finland the Nord Pool is represented by the Finnish power exchange EL-EX. Finally, in 2000 the Nordic market became fully integrated as Denmark joined the exchange. In 2010, Nord Pool Spot was again enlarged as it opened a new bidding area in Estonia.

Although the Nord Pool was built almost at the same time as the original Pool in England and Wales, they were built independently, and as a result they ended in quite different structures. The main differences of the Nord Pool and the original British Pool were (1) the mandatory versus voluntary role of the pool, (2) the way in which the balance between supply and demand is controlled, and (3) the incentive of the reform and ownership structure of the industry. Additional to these three issues the market structure is clearly different in the sense that while there are only a few active market participants in the British market, there are over 300 market participants in the integrated Nordic power market.

The basic characteristic of the Nord Pool is the voluntary participation since in the Nordic power market there is no obligation to buy or sell through the Pool. Instead, also bilateral contracts outside the Nord Pool are accepted. This means that in the Nordic model the real-time dispatch[8] and the merit order[9] dispatch have been strictly separated. The central grid operator determines the real-time dispatch,[10] but the merit order dispatch is determined by the outcome of the hourly spot market.[11] Originally the main reason to create a different institutional framework in the Nordic power market is the fact that around two-third of the power is generated in hydropower plants. Thus, the trade at the spot market is primarily motivated by the need to adjust positions as there appear unexpected variations in supply and demand conditions (see [2]).

The Nord Pool closes everyday at noon when the supply and demand bids are cleared against each other and commitments are made for the delivery of the following day on an hourly basis. The interval between the times the bids are made and the actual trading takes place is at least 12 h. It is significant that both generators and consumers are required to plan to meet all the commitments they have made. Because of the time interval between the bids and the actual delivery, a certain amount of fluctuation in the actual supply and demand is unavoidable compared with the commitments made on the spot market. In order to control the balance a regulation system has been created (see [37]). The market participants can hedge their price through financial contracts and thus manage the possible price risks. Financial contracts are traded through Nasdaq OMX Commodities. There are different types of contracts covering daily, weekly, monthly, quarterly, and annual contracts. The reference price which is used in the financial market is based on the Nord Pool Spot price.

The main motivation in the restructuring of Nordic power markets was not privatization but rather the possibility to improve efficiency. Because the Nordic

---

8 The real-time dispatch refers to the real, implemented, sequence according to which different production units are utilized.

9 The merit order dispatch refers to the sequence, according to which different production units are utilized if cost minimization is used as a crucial argument. In other words, units are organized such that the unit that has the lowest marginal costs is utilized first, the unit that has the second lowest marginal costs is utilized next, and so on.

10 As in the British system.

11 Operated by the independent Nord Pool.

countries provide the first multinational electricity markets, where it is possible for the seller and the buyer to trade between nations, the possibility of congestion in the transmission grid had to be carefully considered and distinctive rules had to be created. Transmission services are based on so-called point tariffs. Generally speaking, this means that at each location there is a given price per unit of power fed into or tapped from the transmission system. This price is independent of the location of the buyer or the generator of that power. The geographical distance between the seller and the buyer does not affect the price of the corresponding transmission service. However, whenever there is congestion in the network prices may vary between countries. Furthermore, Norway can be split into five different price areas, Sweden into four, and Denmark into two. Finland and Estonia are always treated as one price area. The Transmission System Operator (TSO) decides the number of bidding areas and Nord Pool Spot calculates a price for each bidding area for each hour of the following day (see [37]).

2.2.2 Current Structure of the Nordic Power Market

Currently, Nordic countries continue to run a common power exchange, the Nord Pool. The mix of production technologies in the Nordic power market is quite large and it has been argued that it would improve the efficiency of production if market participants could trade between countries.

Nord Pool Spot manages the capacity on the interconnectors between the Nordic countries and the cables that connect the bidding areas in Norway. A privileged place on a bottleneck could be abused by a commercial participant and it is therefore essential that the capacity is given to a neutral party (see, e.g., Nord Pool webpage for the discussion on bottlenecks).

The total net electricity production in the Nordic market was 367 TWh in 2009 (see [12]). Of the produced electricity 72.6 TWh was based on nuclear power, 205.1 TWh on hydropower, 61.7 TWh by using conventional thermal power plants, and 27.6 TWh was based on other renewables. The amount produced by hydropower can change much from year to year depending on precipitation. When the precipitation of the year is low, power is exported from Finland and Demark to hydro-dominated regions and in the high precipitation years the opposite is true. Sometimes the precipitation is so high that some thermal capacity is idle during that period. There are five nuclear power plants currently operating in Nordic countries. Three of them (10 reactors) are located in Sweden whereas two of them (four reactors) are located in Finland. There is also one more 1,600 MW reactor under construction and permission to construct two more reactors in the near future in Finland. Characteristic to the Nordic energy markets is that a large part of the conventional thermal power is produced by combined heat and power (CHP) plants. The peak technology includes oil-fired condensing power plants as well as gas turbines. In our simulations below, we divide our technologies into five representative technology groups based on the main characteristics of the Nordic power market.

Nordic power markets operate under the European Commission's internal emissions trading. At present, the emissions trading only concerns carbon dioxide emissions. The emissions trading scheme is meant to operate so that the emissions of the companies under the scheme keep the predefined total emissions quantity within the limits. For electricity markets, the Emissions Trading Act is applied to carbon dioxide emissions of such power stations for which the thermal input is more than 20 MW and also for the smaller combustion installations connected to the same district-heating network. Typically, the issuance of permits lies with the National Energy Market Authority. The amount of issued permits by power stations is less than their yearly emissions. Power producers can buy extra permits from the emission permit markets. This increases the costs of technologies under emission trade. As the new emission trade period starts at the beginning of 2013 the amount of issued permits per power station clearly reduces and more permits have to be bought from the emission permit market. The impact of this on the price of permits remains to be seen.

The aggregate demand for electricity in Nordic countries has been quite stable from year to year and the increase has been mainly due to economic growth. Some yearly variations happen along with variation in temperature. Price elasticity of demand has typically been very low because the price that final customers face is typically fixed for some period of time and prices do not follow the pattern of wholesale prices in the short run. Economists have argued that the absence of Real-Time Pricing is one of the most obvious shortcomings of the functioning of the electricity markets from an efficiency point of view. This is mainly because if demand is not responding to the prices we need too much reserve power capacity to meet the demand also in the highest peak hours (and this is of course very costly). The inelasticity of demand may also enhance the ability for producers to use market power (see, e.g., [27]).

2.3 Assessment of Deregulation Processes

Prior to the worldwide wave of deregulation, electricity was supplied by regional monopolies that owned both the power plants and the transmission lines for the distribution of power. Some form of regulation was used to set the rate of return of profit for the utilities in all nowadays restructured countries. Although it was recognized over 30 years ago that the character of electricity generation had removed from natural monopoly to the potentially competitive activity, there was no real pressure for the creation of a "deregulated generation market" until the 1990s. This was either because the political atmosphere supported it (as in the United Kingdom), or because large industrial customers did not want to pay vertically integrated traditional utilities for their expensive electricity (as in the U.S.). It is also possible, as argued in public discussion, that big generators started to support restructuring of the electricity industry because they saw the possibility to increase their profits through a speculative market.

There are many observable differences in how the deregulated electricity supply industries can be organized. The interaction between created market rules and the prevalent market structure of the industry determines whether economically efficient prices can be set by these markets (see, e.g., [41, 42]). According to our view, the success of the deregulation process and the target to improve efficiency depends on six issues. First, the *number of active players* in the wholesale market seems to be important, not so much whether the wholesale is carried out through some kind of a spot market or a bilateral market. Second, the *rules of the bidding procedure in the wholesale market* clearly seem to affect the outcome of the market. Third, the *organization of the demand-side operation* in the wholesale market is much more important than has been recognized so far. Fourth, the *transmission grid should offer a neutral market* place for competitive activities. This requires that the access to the transmission grid is based on equality and furthermore that the transmission capacity is high enough to guarantee its efficient operation. Fifth, it seems that some *production technologies* make it easier for companies to use market power than others. And finally, *the ownership structure* may have some effect on the outcome of the market. Next, each of these six issues is discussed.

The first thing that clearly seems to affect the success of the restructuring process is the *number of active players* in the competitive markets. In some countries, such as the Nordic countries, the number of the market participants has been seen to be a crucial issue in order to achieve the target of deregulation, i.e., efficiency improvement. However, some other countries, such as the United Kingdom, have relied on the market performance even when there have been only few active companies in the market. This has been the case, even though already before deregulation Henney [19] argued that the British generating companies should be split at least into nine separate companies. Sykes and Robinson [40] also proposed that there should be at least five or six competing generating companies in the competing electricity market in order to reach the goal of lower prices. Green and Newbery [17] suggested that the generators using thermal power (in the United Kingdom) should be divided into five generators of equal size. Further, they argued that the scope of exercising market power has been considerably underestimated.

Thus, it was not a surprise that very soon after liberalization it became clear that the two major generators in the UK, National Power and PowerGen, had sufficient market power to raise prices in the Pool (see, e.g., [14, 44, 45]). This was possible because of two things, *the structure of bidding procedure* and *the determination of demand* when market price is calculated. The resulting price of the bidding procedure is called the system marginal price (SMP) and it is used in electricity spot markets worldwide. It is based on the bid of the most expensive set in normal use. The system is defined such that the lowest cost generating capacity is dispatched first, unless such dispatch will compromise the system integrity. According to this dispatching procedure "least-cost merit order" gives rise to an upward sloping aggregate electricity supply function for each price period of the system. The SMP is determined combining the expected demand function to this supply function (see [41]).

In an electricity supply system where there are only few large companies they can manipulate the SMP by removing some of their capacity from the market. The generators thus are able to maximize their profits by keeping the industry's capacity at a lower level than would be efficient. It has also been claimed that large generators may bid some of their stations above their marginal costs. As a result these stations will be displaced in the merit order, sacrificing some market share, but in that way the infra-marginal stations can earn more because of the higher level of SMP (see [17]). In markets where there are many active players, as in the Nordic power market, the influence of one market participant on the outcome of the market is smaller than in the case of only few big suppliers, and thus price manipulation is more difficult.

The way in which the *demand function* is constituted has also great influence on the outcome of the market. Demand may be based on the estimation by the system operator or on the true bids of purchasers of electricity (as in the Nordic power market). If demand is based on the estimation, the operators' forecasts for demand can be readily available for generators prior to their submissions of bid prices and availability declarations for the next day. In this kind of a system generators can compute the forecast for demand for all load periods before they submit their bid prices and available generation capacity. Wolak and Patrick [43] argue that this market rule clearly improves the possibilities for generators to exercise market power. In the history, market power has been observed to be clearly a problem at least in the United Kingdom and in the state of California (see [8]). Another demand-side issue which has recently gained more and more importance is *real-time demand responses* to the changing marginal costs of production. Long before worldwide electricity deregulation and restructuring began, it was known that the marginal cost of producing electricity could change significantly according to the time of the day. This means that the true costs of consuming electricity also vary hour by hour. Consequently, economists have argued that retail electricity prices should also fluctuate hour by hour reflecting their true opportunity costs. The problem has been insufficient metering technology. Recently, however, new technology has enabled hour-by-hour measuring of electricity consumption and hence the technology constraint is disappearing (see, e.g., [7]).

A restraint that can significantly distract the operation of competitive markets is *the operation of the transmission grid* (see, e.g., [25]). Transmission grids have been a clear problem in some parts of the United States and also in New Zealand in the history. For example, there are areas in the United States, in which transmission lines become easily congested, which makes free competition difficult. For example, a significant amount of the generation units in California are so-called "must-run" units. This maintains local market power also in the case of free competition between different states. It is possible to diminish the problem of market power so that operators can ignore or cancel the bids made by generators that have been suspected of exercising local market power. However, the best way to improve competitive conditions might be to increase the contestability of separate markets by improving the transmission infrastructure (see [9]).

The *diversity of generation technologies* seems to impact the outcome of the competitive market. It is interesting that, for example, market prices in the markets dominated by fossil fuel technology, for example, in the United Kingdom and in the state of Victoria (Australia), have been much more volatile[12] and also higher than the prices in the markets dominated by hydroelectric or nuclear power capacity, such as the Nordic power market and New Zealand. Possible explanations can be that it is more difficult to manipulate market prices when production is based on so-called must-run technologies. Also, the *ownership structure* may have an influence on the outcome of the market price, since the majority of the generating capacity in the United Kingdom and Victoria is privately owned and thus their objective may be pure profit maximization, whereas for example in the Nordic countries large state-owned generation companies have significant market share and thus their objective may be wider than just profit maximization. Consequently, some of the price volatility in the United Kingdom and Victoria may be explained as episodes of the successful and unsuccessful attempts to exercise market power (see [41]).

Is it, then, possible to draw conclusions about the success or failure of deregulation in general on the basis of the international experiences? Clearly, deregulation has offered some benefits, but it also has some weaknesses. Up to now, it seems that if the deregulation is carried out such that the "accurate" market structure is designed carefully and effective market rules can be created, the deregulation can result in increased efficiency and lower prices. However, it should be noted that we are still far away from a perfectly competitive industry. Thus we can conclude this chapter by the words of John Kay [28]: "the real benefits of competitive markets over central planning are that decisions are made on a smaller scale, and a diversity of views can be implemented. This makes the consequences of good and bad decisions more obvious. Errors can be more quickly corrected, and the expectation that individuals may be held responsible for the outcome helps judicious decision-making. Markets are not a perfect form of economic organization. They are just better than the alternatives."

References

1. Alchian A (1965) Some economics of property rights. II Politico 30:816–829
2. Amundsen E, Bergman K (2000) Will cross-ownership reestablish market power in the nordic power market? Working paper no. 56/00, SNF, Bergen, Norway
3. Amundsen E, Nesse A, Tjøtta S (1999) Deregulation of the Nordic power market and environmental policy. Energy Econ 21:417–434

[12] See Wolak [42] for evidence and more detailed discussion of volatile prices.

4. Averch H, Johnson L (1962) Behavior of the firm under regulatory constraint. Am Econ Rev 52:1053–1069
5. Baron D, Myerson R (1982) Regulating a monopolist with unknown costs. Econometrica 50:911–930
6. Baumol W, Panzar J, Willig R (1982) Contestable markets and the theory of industry structure. Harcourt Brace Jovanovich, San Diego
7. Borenstein S (2009) Electricity pricing that reflect its real-time cost, NBER reporter, Winter, 9–12
8. Borenstein S, Bushnel J, Wolak F (2000) Diagnosing market power in California's restructured wholesale electricity market, NBER working paper series, no. 7868
9. Borenstein S, Bushnell J (1999) An empirical analysis of the potential for market power in California's electricity industry. J Ind Econ 47:285–323
10. Bushnell J, Wolfram C (2005) Ownership change, incentives and plant efficiency: the divestiture of U.S. electric generation plants, CSEM working paper no. 140
11. Derthich M, Quirk P (1985) The politics of deregulation. The Brookings Institution, Washington
12. Eurelectric (2010) Power statistics. www.eurelectric.org. Accessed 10 Aug 2011
13. Fabrizio KR, Rose NL, Wolfram CD (2007) Do markets reduce costs? Assessing the impact of regulatory restructuring on U.S. electric generation efficiency. Am Econ Rev 97:1250–1277
14. Green R (1994) Britain's unregulated electricity pool. In: Einhorn M (ed) From regulation to competition; new frontiers on electricity markets. Kluwer Academic Publisher, Boston
15. Green R (2006) Electricity liberalization in Europe—how competitive will it be? Energy Policy 34:2532–2541
16. Green R (2007) EU regulation and competition policy among the energy utilities, discussion papers 07-02, Department of E conomics, University of Birmingham
17. Green R, Newbery D (1992) Competition in the British electricity spot market. J Political Econ 100:929–953
18. Grohnheit P, Olsen OJ (2002) Organization and regulation of the electricity supply industry in Denmark. In: De Paoli L (ed) The electricity industry in transition. Organization, regulation and ownership in EU member states. Franco Angeli, Milano, pp 123–162
19. Henney A (1987) Privatise power: restructuring the electricity supply industry. Policy study no. 83. Centre policy studies, London
20. Hjalmarsson L (1996) From club-regulation to market competition in the Scandinavian electricity supply industry. In: Gilbert RJ, Kahn EP (eds) International comparisons of electricity regulation. Cambridge University Press, New York, pp 126–178
21. Joskow P (1987) Productivity growth and technical change in the generation of electricity. Energy J 8:17–38
22. Joskow P (1998) Electricity sectors in transition. Energy J 19:25–52
23. Joskow P (2008) Incentive regulation and its application to electricity networks. Rev Netw Econ 7:547–560
24. Joskow P (2008) Lessons learned from electricity market liberalization. Energy J 29:9–42
25. Joskow P, Tirole J (2000) Transmission rights and market power on electric power networks. RAND J Econ 31:450–487
26. Joskow P (2010) Market imperfections versus regulatory imperfections. CESifo DICE Rep 8(3):3–7
27. Kauppi O (2009) A model of imperfect dynamic competition in the Nordic power market, Acta Universitatis Oeconomicae Helsingiensis A-350. Dissertation, Helsinki School of Economics
28. Kay J (2001) Financial times. 18 April 2001
29. Kopsakangas-Savolainen M, Svento R (2010) Comparing welfare effects of different regulation schemes: an application to the electricity distribution industry. Energy Policy 38:7370–7399

30. Laffont J-J, Tirole J (1986) Using cost observation to regulate firms. J Political Econ 94:614–641
31. Laffont J-J, Tirole J (1993) A theory of incentives in regulation and procurement. MIT Press, Cambridge
32. Laffont J-J, Tirole J (1994) Access pricing and competition. Eur Econ Rev 38:1673–1710
33. Newbery D (1997) Privatization and liberalization of network utilities. Eur Econ Rev 41:357–383
34. Newbery D, Green R (1996) Regulation, public ownership and privatization of the English electricity industry. In: Gilbert RJ, Kahn EP (eds) International comparisons of electricity regulation. Cambridge University Press, New York, pp 25–81
35. Newbery D, Pollitt M (1997) The restructuring and privatization of Britain's CEGB—was it worth it? J Ind Econ XLV:269–303
36. Niskanen W (1968) The peculiar economics of bureaucracy. Am Econ Rev, Pap Proc 58:298–305
37. Nord Pool Spot (2011) The Nordic electricity exchange and the Nordic model for a liberalized electricity market. http://www.nordpoolspot.com/Global/Download%20Center/The-Nordic-Electricity-Exchange-and-the-Nordic-model-for-a-liberalised-electricity-market.pdf. Accessed 24 Nov 2011
38. Pollitt M (1995) Ownership and performance in electric utilities, Oxford Institute for Energy Studies. Oxford University Press, Oxford
39. Pollitt M (1997) The impact of liberalization on the performance of the electricity supply industry: an international survey. J Energy Literature III:3–31
40. Sykes A, Robinson C (1987) Current choices: good ways and bad to privatise electricity. Policy study no. 87. Centre Policy Studies, London
41. Wolak F (2001) Market design and price behavior in restructured electricity markets: an international comparisons. Working paper http://www-leland.stanford.edu/~wolak. Accessed 9 Sept 2001
42. Wolak F (2005) Lessons from international experience with electricity market monitoring. World bank policy research working paper 3692
43. Wolak F, Patrick RH (1997) The impact of market rules and market structure on the price determination process in the england and wales electricity market. http://www.stanford.edu/group/fwolak/cgi-bin/?q=node/3. Accessed 5 Sept 2011
44. Wolfram C (1998) Strategic bidding in a multi-unit auction: an empirical analysis of bids to supply electricity in England and wales. RAND J Econ 29:703–725
45. Wolfram C (1999) Measuring duopoly power in the British electricity spot market. Am Econ Rev 89:805–826

Chapter 3
Modeling Energy Production System

3.1 Background

There are various ways to model the energy production system. Our approach in the modeling is forward looking in the sense that we want to include into the model the main characteristics of the future energy markets. Consequently, we assume that the demand side of the energy system is at least partly responsive to the real-time pricing (RTP) of the market. The basic structure of the Borenstein and Holland model [2] and Borenstein [1] model includes desirable features of this kind of a structure.

In this chapter we describe and construct a basic energy market model which we use in the simulation applications in Chaps. 4–6. Simulation applications relate to Nordic power markets and accordingly, also the basic model structure is constructed by using the Nordic power market production structure as a starting point. The following model can be, however, generalized to any other kind of market structure as well.

3.2 Structure of the Model

Following Borenstein and Holland [2] and Borenstein [1] assume a constant elasticity demand function. The demand side of the model is based on calibration of the demand function to correspond to real, market-specific demand profiles. Calibration is done by calculating an "anchor point" A_h for each hour of the year. In order to specify the anchor point we need to assume some constant market price. In what follows we use a price that would allow producers just to break even if they were charging that constant price from all customers. See Borenstein [1] for a more detailed discussion of the role of this constant price. The anchor point can be determined as follows: $A_h = D_h/p_c^{\varepsilon}$, where D = demand,

M. Kopsakangas-Savolainen and R. Svento, *Modern Energy Markets*,
Green Energy and Technology, DOI: 10.1007/978-1-4471-2972-1_3,

h = hour, and p_c^{ε} is the constant price. Notably, it is more important to have accurate information about the shape of the hourly demand distribution than very specific information about the constant price. These anchor points are used to scale the demands to the load curve.[1] Now let α, $0 < \alpha \leq 1$, be the share of customers on RTP, ε the price elasticity for homogeneous customers, and p_r the retail price of electricity for the RTP customers and p_f is the flat rate price for non-RTP customers. The demand for electricity for hour h can then be written as:

$$D_h\left(p_r, p_f\right) = \left[\alpha p_r^{\varepsilon} + (1-\alpha) p_f^{\varepsilon}\right] A_h, h = 1, \ldots, 8760. \tag{3.1}$$

In the RTP simulations α and ε are varied exogenously. We assume a competitive market structure in generation and in retail. We also assume that both retailers and generators will maximize profits and their profit functions can be written as follows:

$$\pi_R = \sum_{h=1}^{8760} \left[\left(p_f - w_h\right)(1-\alpha) D_h\left(p_f\right) + \left(p_{\mathrm{r}} - w_h\right)\alpha D_h\left(p_{\mathrm{r}}\right)\right], \tag{3.2}$$

$$\pi_G = \sum_{h=1}^{8760} \left(w_h D_h - c D_h\right) - rK, \tag{3.3}$$

where R = retail sector, G = generation sector, w = wholesale price, c = marginal generation costs, and rK = annual capital costs.

From Eqs. (3.2) and (3.3) we can solve the short- and long-run equilibria of the generation and consumption systems. The logic of the simulation system is that the whole generation system is constructed MW by MW under the assumed economic principles (i.e., generators maximize profits according to Eq. 3.3). After the construction of the whole generation system we turn to the retail sector and adjust the flat rate until the profits (see Eq. 3.2) for the retail sector are equal to zero. Then we turn again to the generator sector and rebuild the whole system MW by MW with this new flat rate. These two rounds are repeated until we reach the long-run equilibrium of the whole system.

3.2.1 Equilibrium with One Technology

To describe the simulation model and the solution procedure in a simple way assume first that only one capacity type (peaker capacity) without any scale effects is constructed and used. The short-run supply curve with an assumed short-run capacity $K_{\mathrm{tot}}^{\mathrm{SR}}$ is thus an inverted L-shaped curve as depicted in Fig. 3.1.

---

[1] See Fig. 4.3 for the description of the load curve.

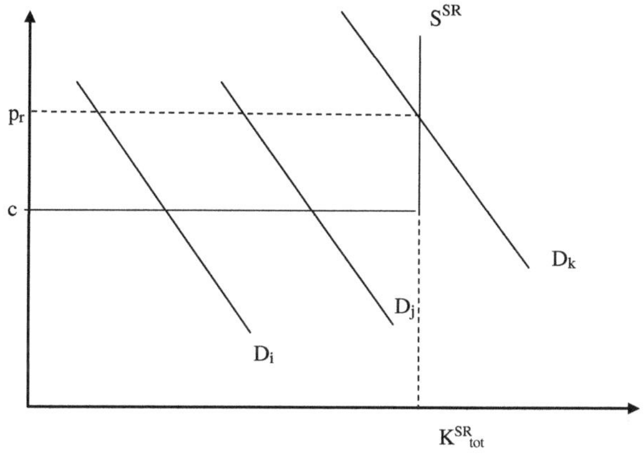

Fig. 3.1 The energy system with only one technology

The short-run profits by hour are maximized by maximizing $\pi_G^{SR} = w_h D_h - cD_h$ with respect to given demand so that we have the known result: $w_h = c$. But this is true only when demand is not greater than the capacity limit, i.e., demands D_i and D_j in Fig. 3.1. For those hours when demand exceeds in the short-run capacity limit, the RTP must adjust to take care of market clearing. In equilibrium it has to be that demand equals supply, i.e.:

$$D_k = \left[\alpha p_r^{\varepsilon} + (1-\alpha)p_f^{\varepsilon}\right]A_h = K_{tot}^{SR} \tag{3.4}$$

From where we can solve for the price p_r:

$$p_r = \left[\frac{K_{tot}^{SR} - (1-\alpha)p_f^{\varepsilon}A_h}{\alpha A_h}\right]^{-\varepsilon} \tag{3.5}$$

So the wholesale pricing logic is as follows:

$$w_h = c \quad \text{when } D_h \leq K_{tot}^{SR} \tag{3.6}$$

$$w_h = p_r \quad \text{when } D_h > K_{tot}^{SR}. \tag{3.7}$$

Once we know the demand and wholesale price for each hour we can calculate profits for the generators. Because of competition it is assumed that capacity is built to the point where profits are equal to zero.

We still have to solve for the flat rate. Competition forces the retail sector to zero profits also in the short-run. Abstracting from transmission and retail costs the real-time retail price must always be equal to the wholesale price. Should this not be the case there would be possibilities for undercutting the market price and this will go on as long as the retail price exceeds the wholesale price. So the zero profit condition for the retail sector reduces to:

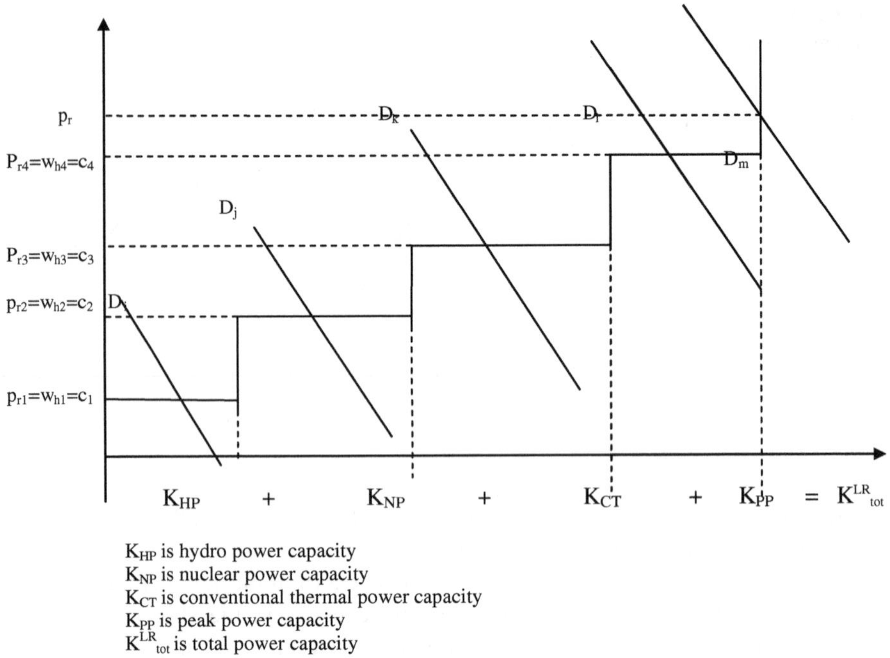

Fig. 3.2 The long-run structure of the energy system

$$(1-\alpha)\sum_{h=1}^{H}\left(p_f^{\varepsilon}-w_h\right)D_h\left(p_f\right)=0 \tag{3.8}$$

from which we can solve for p_f:

$$p_f=\frac{\sum_h^H w_h D_h\left(p_f\right)}{\sum_h^H D_h\left(p_f\right)}, \tag{3.9}$$

i.e., the zero profit short-run flat rate is a weighted average of the real-time wholesale price with weights being the relative quantities demanded by customers facing a flat retail price.

In the long run, capacity is built to the point where both generators and retailers receive zero profits. As shown in Borenstein and Holland [2] and Borenstein [1] this kind of mechanism leads to a unique long-run equilibrium for the total generation capacity, $K_{\text{tot}}^{\text{LR}}$.

3.2.2 Equilibrium with a Technology Mix

We can use the same kind of procedure as explained above to solve for the technology mix. Now, however, the total technology built consists of different types of technologies with different capital and variable costs. In the following we have used representative technology mix from the Nordic power market (see [3]). Technologies used are divided into four categories from which two, peak and mid-merit technologies, are freely optimized and two, hydro and nuclear technologies, are capacity constrained. Because of competition those technologies that are not capacity constrained are built to the point where the profits equal zero. The technologies that are capacity constrained are built either to the point where their profits are zero or to the limit of the capacity constraint. In practice, we first solve for short-run total capacity $K_{\text{tot}}^{\text{SR}}$ by using peaker capacity only (as described above). Then we start to replace peaker capacity with mid-merit capacity. We construct the mid-merit capacity to the point where the short-run profits for mid-merit capacity go to zero. Finally, we construct the capacity constrained technologies either to the point where their short-run profits are zero or to the point where they reach the limit of their capacity.

After the short-run technology mix for generation is solved we turn to the retail market and solve the short-run equilibrium for the retail market. Short-run equilibrium for the retail market is reached when the retail sector produces zero profits (see Eq. 3.8). From the zero profit condition we can solve the new flat rate. With this new flat rate we start to construct again the whole generation technology mix. These two rounds are repeated until the equilibrium conditions for both sectors are simultaneously fulfilled. The long-run energy system equilibrium looks like the system in Fig. 3.2.

To compare and interpret the simulations in the following sections we also need a baseline solution with no customers on RTP. In order to meet the demand in this case, in all hours there must be sufficient capacity so that the market clears on the supply side. This means that the wholesale price cannot be higher than the marginal generating cost of the technology with the highest marginal cost. This requires an additional wholesale payment to generators in order to assure that the market clears from the supply side and that the generators have enough revenue to cover their fixed costs. As Borenstein [1] shows this payment must be equal to the annual fixed costs of a unit of peaker capacity. In the baseline solution this payment is made to all generators in order not to distort the mix of capacity.

3.3 Description of the Algorithm

Next we give the summary of the basic algorithm used in the simulations in Chaps. 4–6.

a) Make an assumption for α and ε. Begin with an initial guess for the peaker technology capacity.[2] Then expand the quantity of peaker capacity, combine this short-run supply function with the hourly demand and calculate short-run

[2] We begin with peaker technology because if used at all it will be used in the highest demand hour.

profits of the peak power generator. If profits are positive expand the quantity of peaker capacity again, recalculate the profits, and so on. Continue the expansion of the capacity until expansion by one more unit causes profits to go negative for peak power generators. The resulting amount of capacity will typically be the short-run equilibrium of total amount of all capacities (K_{tot}^{SR}). Note that in this equilibrium where only peak technology is used, all profits to the generator are earned when the production is equal to the total amount of peak capacity. In hours where equilibrium quantity is less than K_{tot}^{SR}, price is equal to the marginal costs and profits resulting from these hours are equal to zero (see Fig. 3.2).

b) Next, start substituting peaker capacity with mid-merit capacity. The mid-merit capacity will be lower on the supply function than the peaker capacity and thus it will be used in all hours before any of the peaker capacity. Follow the same procedure as with the peaker capacity, namely: expand mid-merit capacity, combine short-run supply function with the demand, and calculate profits of the mid-merit units. If profits are positive continue expansion until expansion of the mid-merit capacity by one more unit will cause profits of all mid-merit units to go negative. We refer to the total amount of mid-merit capacity that still results in positive profits by K_{mid}.
c) Next begin substituting mid-merit capacity by the baseload capacity. Follow the same procedure as in stages a and b above. Note, however, that we have assumed that the baseload capacity is constrained.[3] Expand the baseload capacity, combine the new short-run supply function to the demand, and calculate profits. Again, if profits are positive continue expansion. Expand the capacity up to the point where expansion of capacity by one more unit will cause profits of all baseload units to go negative or where the expanded capacity reaches the capacity constraint. Since the baseload capacity in our case refers to nuclear power, the capacity constraint is at a lower level than it would be if the amount of the capacity would be determined by free optimization. Refer to the amount of baseload capacity by K_b.
d) Next, take the capacity of hydropower generation into account at each hour of the year. Refer to the amount of hydropower capacity by K_{HP}.
e) Now the capacities of all types of technologies have been determined. Short-run amounts of capacities can be determined by using the following recursive logic:

   i. Hydropower: $K_{HP} = K_{HP}$
   ii. Nuclear power: $K_{NP} = K_b - K_{HP}$
   iii. Conventional thermal power: $K_{CT} = K_{mid} - K_{NP} - K_{HP}$
   iv. Peaker power: $K_{PP} = K_{tot}^{LR} - K_{CT} - K_{NP} - K_{HP}$

[3] Although the baseload capacity refers to the nuclear power, for modeling reasons we first assume that the baseload constraint is equal to the capacity of nuclear power and hydropower together.

f) The final step of this procedure is to note that this equilibrium may not satisfy the retailers' breakeven condition. So one must calculate the profits that the retailers earn from flat rate customers in this specific wholesale-producers equilibrium (note that profits for the retailers from RTP customers are always equal to zero). If the retailer profits from flat rate customers are negative, one adjusts the flat rate price up and if profits are positive one adjusts the flat rate price down. After this adjustment we need to re-simulate the capacity so we go back to point a. This procedure has to be continued until we reach equilibrium where retail markets also yield zero profits. This is the long-run unique competitive equilibrium energy system size and structure for a given set of available technologies, technology constraints, and share of customers on RTP and on flat rate.

References

1. Borenstein S (2005) The long-run efficiency of real-time electricity pricing. Energy J 26:1–24
2. Borenstein S, Holland S (2005) On the efficiency of competitive electricity markets with time-invariant retail prices. RAND J Econ 36:469–493
3. Kopsakangas-Savolainen M, Svento R (2011) Real-time pricing in the nordic power markets. Energy Econ. doi:10.1016/j.eneco.2011.10.006

Chapter 4
Real-Time Pricing; An Application to the Nordic Power Markets

4.1 Background

Long before worldwide electricity deregulation and restructuring began it was known that the marginal cost of producing electricity could change significantly according to the time of the day. This means that the true costs of consuming electricity vary also hour by hour. Consequently economists have argued that retail electricity prices should also fluctuate hour by hour reflecting their true opportunity costs. The problem has been insufficient metering technology. Recently, however, new technology has enabled hour-by-hour measuring of electricity consumption and hence the technology constraint is disappearing.

Since 1990 some electricity producers (in the USA and in Europe) have signed real-time pricing (RTP) contracts with big industrial customers. Only recently, after the introduction of the new metering technology, large-scale RTP has become a true possibility for electricity contracts. The studies on RTP have been concentrated on the impacts of short-run pricing and long-run investment inefficiencies (see, e.g., [6]), on the incentives of the seller to exercise market power (see, e.g., [7, 8, 13]) and on the impacts to the wealth transfer (see, e.g,. [4]). According to Borenstein and Holland [6], the potential social gains from switching to RTP are almost certainly many times greater than the estimated costs of implementing such a program.

The result that RTP clearly offers potential efficiency gains is often confused with energy savings and the ability to reduce greenhouse gases. RTP might in some cases decrease consumption and in some cases work in the opposite direction. The same is true for the greenhouse gases. This stresses the importance of research if RTP is going to be implemented on a large scale.

It has been argued (see, e.g., [12]) that the effect on the environment of redistributing consumption from peak to off-peak periods depends, among other things, on the production structure of the particular area. Holland and Mansur [12] argue that when coal provides the base load power and gas-fired stations are used in peak load periods, RTP is likely to increase greenhouse gases, but for those

M. Kopsakangas-Savolainen and R. Svento, *Modern Energy Markets*, Green Energy and Technology, DOI: 10.1007/978-1-4471-2972-1_4,

systems which rely less on coal, consumption smoothing as a result on RTP can potentially benefit the environment.

As consumption is not expected to diminish in the Nordic power markets and at the same time some older production technologies are going to retire, we are in a situation where the capacity is becoming more limited in peak demand periods. RTP would likely decrease the variance of load but there is uncertainty of what would happen to the emissions. From the investment efficiency point of view, it is important that we motivate electricity consumers to save energy during peak demand periods so that those periods can be avoided whenever possible.

A key question related to the effectiveness of RTP concerns the possibilities that the customers have in responding to price changes. There exists a body of the literature trying to estimate the demand elasticities specifically for RTP customers. The problem is that the data sets are in many cases rather small and most contracts on RTP have in fact been offered to industrial customers. The core studies in this literature are Herriges et al. [11], Patrick and Wolak [17], and Schwarz et al. [18]. These studies use a variety of models but the general picture seems to be quite consistent: the daily and hourly elasticities are rather small by absolute value, varying between −0.01 and −0.10, but still big enough to drive considerable changes in peak loads. The customers seem to be heterogeneous in that they reduce demand by varying amounts during peak price periods. Also, the responses seem to link together with the absolute level of the price, and time-related learning has been detected with dynamic models.

A much less studied question is, what effects does a change to an RTP system have on the energy generation structure? Is it possible to get rid of the most expensive peaker power technologies? Borenstein and Holland [6] and Borenstein [2–5] have studied the impacts of RTP on the whole energy sector. In their 2005 paper, Borenstein and Holland studied the short-run pricing and long-run investment inefficiencies resulting from a system, where some or all customers face retail prices that do not vary with the wholesale market price. These papers are based on theoretical modeling and they show that the deregulated competitive equilibrium without RTP is inefficient. This inefficiency leads to inefficient investment levels. When only a part of customers is in RTP they produce an externality to flat rate customers because the wholesale price and investment levels are not optimal.

Borenstein and Holland also used simulations in these papers to examine the social gains from switching to RTP. The results are "significant and at the same time sobering" as they put it [2]. The significant part comes from the results, which show that even though the social gains are not that big, they almost certainly are many times greater than the expected costs of implementing an RTP switching program. The sobering part comes from the results showing that moving to theoretically sound marginal cost-based pricing produces social gains that are likely to be only 5% or less of the energy bill. The reason for this is that an electric system must always stand ready to meet all demand at the retail price and a detriment of the flat rate pricing mechanism is the need to hold substantial capacity that is hardly ever used. Utilities optimize against this restriction by building peaker capacities that have low capital costs and high operating costs. The costs to society of this idle capacity turn out to be not as great as one might expect. Also substantial elasticities would be needed to eliminate the need for these capacities.

Faruqui et al. [10] have simulated the effects of dynamic electricity pricing in California. They have calculated an hourly RTP that would compare to a flat rate. Using the simulated real-time profile in a modeling system called The PRISM (Pricing Impact Simulation Model) they found that dynamic pricing rates have the potential to reduce system peak demands between 1 and 9%, with the variation in the magnitude of demand response coming primarily from the offered rates and how they are deployed. The potential benefit to California from the deployment of dynamic pricing is valued at $0.6 billion at the low end and $6.0 billion at the high end.

Recently, according to the US program where residential consumers were on RTP, Allcott [1] finds that households are statistically significantly price elastic, conserving energy during peak hours. Remarkably households did not increase average consumption during off-peak times.

The main purpose of this chapter is to look at the effects RTP has on the need for long-run investments in the Nordic power markets and economic welfare (see [14]). We are also interested in what kind of impact the inclusion of an emission permit market has on the RTP results. We follow the methodology based on Borenstein [5] and Borenstein and Holland [6] model and presented in Chap. 3 in our simulations. In this chapter we contribute to the Borenstein and Holland [6] and Borenstein [5] studies as follows. Firstly we assume that some technologies—hydro and nuclear—are capacity constrained. Secondly we apply the model to the real market context. Thirdly we analyze the impact of emission trade (ET) on the results of RTP and show how RTP and ET must be seen as two complementary policy instruments in reaching the energy system efficiencies.[1]

The main contribution of this chapter is that when some customers accept RTP contracts total equilibrium capacity clearly diminishes even when demand is very inelastic (−0.025). This mainly results from the reduction of the peaker capacity if there is no ET and from the reduction of both mid-merit and peaker capacity if we assume simultaneous ET. This result is strengthened when demand becomes more elastic and more customers move to the RTP group. Another clear observation is that price falls during the last peak demand hour with more elastic demand and more customers on the RTP. The impact of RTP on the total annual consumption is dependent on whether the RTP is implemented with or without ET. With ET it is possible to see energy conservation if demand is price elastic enough. This does not happen without ET.

Also the impact on the billing costs of customers is sensitive to the ET. Billing costs for the customers decrease with all elasticity values and shares of the customers belonging to the RTP group if there is simultaneously ET. This is not true without the ET. This is mainly due to the decrease in consumption and partly due to the decrease in MWh price for RTP customers.

---

[1] In Chap. 6 we give a compact history of the European Union Emission Trading Scheme.

In our simulations those units which are using capacity constrained technologies are able to make profits. According to our results these profits are not very sensitive to the different demand elasticities or RTP scenarios but they are clearly increasing with the ET. This means that hydro and nuclear power producers make quite significant windfall profits and clearly benefit from the ET.

4.2 Data for Simulations

For the simulations we need data for the load profile and cost information of the different production technologies. The production specific cost data is presented in Table 4.1. The variable costs of mid-merit and peaker units are determined with and without ET. Assumed emission price is 23 €/tCO_2. The variable costs without ET are the numbers in the brackets in Table 4.1. Of course the actual mixture of production technologies is more diversified but for the purpose of this analysis somewhat less diversified production structure is sufficient.

Hydropower is a very important technology in the Nordic power market. By nature the hydropower differs from other technologies because of very low variable costs and its dependence on the precipitation of the year. It cannot be clearly identified as belonging to any of the groups: baseload, mid-merit, or peaker capacity. Part of it can be used as baseload capacity but most of it is used as balancing power. The cost of hydropower production of course depends on the type of the power plant and on other environmental factors. The annual capital cost of each technology is calculated by assuming 3% interest rate and by using a standard annuity formula. It is important to use annual capital cost since in the simulation model the demand is determined by the annual hourly load duration curve. Thus, we also need costs determined on a yearly basis. For hydropower we assume that investment costs are 2,000 € per installed kW and that economic lifetime of that capacity is 75 years. Using an interest rate of 3% and the standard annuity formula we end up with the 67,336 € capital cost per year. Variable costs include fuel costs, operation and maintenance costs, and the impact of emission permits' price on the variable costs in the case when we assume ET. In the case of hydropower, variable costs are basically equal to maintenance costs because fuel costs are equal to zero. It is notable that in the simulations we are assuming that the average utilization rate of the hydropower capacity is slightly below 50%.

In simulations nuclear power production is assumed to be baseload production. Our cost data for nuclear power is based on the calculations by Tarjanne and Kivistö [19]. According to their calculations, the specific investment cost of nuclear power is 2,750 €/kW. If we assume a 40 year economic lifetime we end up with the annual capital costs of 118,697 €/MW. It is notable that estimations of the investment costs of the new nuclear power plants vary quite much. For example, the investment costs of the power plant under construction in Finland (Olkiluoto 3) with installed capacity of 1,600 MW is assumed to vary from 3 to 4.5 billion Euros and the expected economic lifetime is up to 60 years.

Table 4.1 Capacity and generation costs

Generation type	Annual capital costs (€/MW)	Variable costs (€/MWh)
Hydropower	67,336	4
Nuclear power (baseload)	118,697	15
Conventional thermal power (mid-merit)	74,657	52 (32.8)
Peaker power	40,200	60 (45)

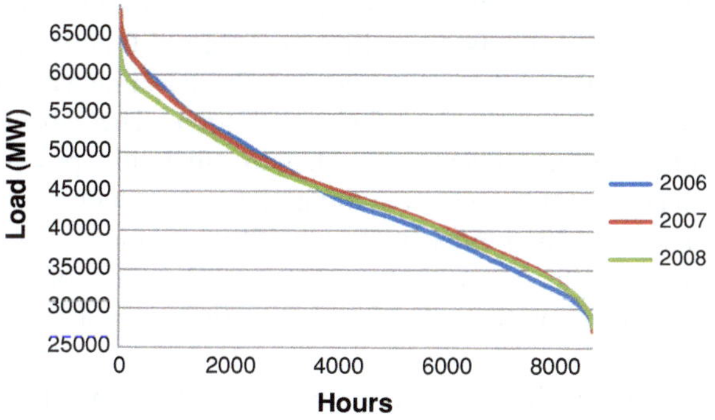

Fig. 4.1 Load duration curves in Nordic power markets from 2006 to 2008

The representative mid-merit technology in our simulations is assumed to be coal and peat power plants. We have again based our costs on the results of Tarjanne and Kivistö [19]. The costs of the representative mid-merit technology are partly (about 35%) based on peat fired conventional thermal production and partly (about 65%) on coal fired production. Specific investment costs of the mid-merit technology are assumed to be about 1,300 €/kW which results in annual capital costs of 74,657 MW over a 25 year economic lifetime.

The capital and variable costs of peaker capacity vary depending on the technology used. Part of the peaker demand can be satisfied by oil-fired condensing power plants, part by older power plants kept as reserve capacity, and the rest by the use of gas turbines. We assume that a representative peaker capacity investment cost is 700 € per installed kW which results in annual capital costs of 40,200 €/MW. If we assume smaller capital costs and higher variable costs it strengthens the impact of RTP in simulations.

In the simulations we need information on the load profile. Our hourly load profile is based on inspecting the hourly consumption data from Nord Pool Spot from 2006 to 2008 (see [15] and [16]). The hourly demand varies from 27,173 MW (lowest hour in 2006) to the 68,111 MW (highest hour in 2007). Because the shapes of the load duration curves are rather similar among different years we have chosen to use the shape of the 2008 load duration curve[2] in our simulations (see Fig. 4.1 for the load duration curves of different years).

[2] We have chosen 2008 load duration curve because the difference among highest and lowest demand hours was smallest. If we were to use a steeper load curve it would only strengthen the following RTP results.

4.3 Results

In Tables 4.2, 4.3, 4.4, 4.5, 4.6 and 4.7 we present our simulation results. The effects of RTP are presented in Tables 4.2 and 4.3. The first table assumes that there will be no ET in the Nordic power market. The second table assumes RTP is introduced into the situation where there is ET, 23 €/tCO_2, in the Nordic power markets. In Tables 4.4 and 4.5 we compare the structural effects of RTP and different emission permit prices. After presenting the basic results of the RTP we present its effects on welfare in Tables 4.6 and 4.7.

In Table 4.2 we first present the baseline values which express the situation when all customers are facing a flat rate. The capacity values presented in the first line present capacities needed to efficiently provide electricity for the demand under flat rate and assuming the cost structure presented above in Table 4.1. The next lines present the equilibrium prices, capacities, and other information under different scenarios of price elasticity of demand and the proportion of customers on RTP. As already mentioned we assume that the hydropower capacity is constrained to the level of 47,816 MW, of which on average 21,899 MW is in use. Only this average amount appears in our total equilibrium capacity values presented in the table below. Nuclear capacity is constrained to the level of 11,636 MW.[3] Notable is that if we assume that some customers are willing to accept RTP contracts the required total equilibrium capacity clearly diminishes even when demand is very inelastic (−0.025). This is mainly resulting from the reduction of the peaker capacity from 15,318 to 12,799 MW. This result is strengthened when demand is more elastic and more customers are moving to the RTP group.

Another clear observation is that price for the last peak demand hour diminishes clearly with more elastic demand and the more customers are on RTP. Also the amount of hours when peak quantity is used changes clearly with demand elasticity and customers on RTP. When demand elasticity is only −0.025 with the share of RTP customers 0.333 the hours at peak quantity are only 60. This means that some peaker units are built just to be in operation 60 h per year and rest of the hours, i.e., 8,700 h they remain unused. The hours of peak quantity in use increase if demand is more elastic, share of the RTP customers increases, or if both happen. If demand is more elastic (−0.3) but the share of the RTP customers remains the same (0.333) the hours at peak quantity on operation increases to 1,081 h per year. A similar effect happens if demand elasticity does not change (remaining at −0.025) but the share of RTP customers increases to 0.999. Now the hours at peak quantity increases from 55 to 351. See Figs. 4.2 and 4.3 for illustrations of the impact of price elasticity and share on the RTP to the load duration curve.

[3] The amounts of generating capacities of hydro and nuclear power are obtained from statistics of Eurelectric [9]. See http://www.eurelectric.org/Statistics.

Table 4.2 Effect of the RTP, without emission trade

Elasticity	Share on RTP	Total annual energy consumed (TWh)	Flat rate (€/MWh)	Mid-merit power (MW) (coal and peat)	Peaker power (MW)	Total equilibrium capacity (MW)	Peak price (€/MWh)	Hours at peak quantity (at 8760)
Baseline scenario		389.38	43.67	14,351	15,318	63,204	–	–
−0.025	0.333	390.08	43.09	14,419	12,799	60,753	6468.40	60
−0.025	0.666	390.51	43.06	14,451	11,277	59,263	2257.90	192
−0.025	0.999	390.83	43.05	14,486	10,260	58,281	1130.13	351
−0.1	0.333	393.90	43.16	14,869	9,723	58,127	894.75	460
−0.1	0.666	394.79	43.20	15,033	7,513	56,081	309.81	857
−0.1	0.999	395.49	43.27	15,213	5,900	54,648	199.50	1,148
−0.3	0.333	402.90	43.41	16,088	6,329	55,952	249.69	1,081
−0.3	0.666	404.97	43.43	16,679	2,599	52,813	127.85	1,775
−0.3	0.999	406.51	43.54	16,967	0	50,502	99.58	3,780

Table 4.3 Effect of the RTP with emission trade

Elasti-city	Share on RTP	Total annual energy consumed (TWh)	Flat rate (€/MWh)	Mid-merit power (MW) (coal and peat)	Peak power (MW)	Total equilibrium capacity (MW)	Peak price (€/MWh)	Hours at peak quantity (at 8760)
Baseline scenario		386.11	61.22	10,206	18,932	62,673	–	–
−0.025	0.333	386.60	60.88	10,246	16,604	60,385	6545.63	55
−0.025	0.666	386.90	60.92	10,289	15,210	59,034	2371.97	164
−0.025	0.999	387.11	60.97	10,316	14,306	58,157	1230.29	298
−0.1	0.333	380.44	60.63	9,533	13,524	56,592	963.55	391
−0.1	0.666	380.72	61.00	9,638	11,709	54,882	355.85	762
−0.1	0.999	380.81	61.31	9,763	10,442	53,740	235.83	1,001
−0.3	0.333	364.55	60.33	7,728	10,165	51,428	282.39	952
−0.3	0.666	363.26	61.23	7,992	7,470	48,997	154.13	1,581
−0.3	0.999	362.74	61.66	8,371	5,498	47,404	122.95	2,008

From Fig. 4.2, we clearly see that when more customers are moving to the RTP group the amount of total capacity clearly diminishes and the amount of peak demand hours increases. Another observation is that with higher share of customers on RTP demand increases somewhat in most of the hours, i.e., in those hours when demand is not satisfied by using peaker capacity. From Fig. 4.3 we see the impact of demand elasticity to the shape of load duration curve. When demand becomes more elastic it clearly increases in lower demand hours but decreases in the peaker demand hours.

The flat parts of the load duration curve happen when there is a technology limit which causes price increase, and during those technology limit hours capacity remains the same. After these hours we move to the next capacity on the system marginal curve (see also Fig. 3.2 above to illustrate the technology limits). This impact is strengthened as demand becomes more elastic.

It is important to note the role of the increase in peak demand hours. When the number of peak demand hours is very low the capacity cost per hour is higher than in the case when the number of peak demand hours is higher. In the case when peak demand hours become sufficiently high it is possible to change the peak demand technology to the technology with higher capital costs but lower marginal costs. One realization of the low number of peak demand hours is the high wholesale price of peak hours.

Annual energy consumption is 389.38 TWh in the baseline scenario, i.e., when all customers are paying a flat rate for their electricity. When demand becomes more elastic or more customers are in the RTP group the total annual energy consumption increases somewhat. One explanation for this behavior is that when demand becomes more elastic RTP customers increase their consumption at the lower demand hours (i.e., when RTP is relatively low) and decrease their consumption at the higher demand hours (i.e., when RTP is relatively high). The impact of higher demand hours increases when the share of the RTP customers increases. Another explanation comes from changes in the flat rate. The flat rate slightly increases as more customers are on RTP and as demand becomes more elastic. Consequently customers who face flat rates decrease their consumption.

Next we discuss on our RTP results when there is ET in the Nordic power market. The main aim is to look at how the general results of RTP change with the ET. We still assume that hydropower and nuclear power capacities are constrained as in the first simulation. The demand elasticity and RTP scenarios are similar to those above.

From the results we see some clear differences to those presented in Table 4.2. First observation to note is that the flat rate clearly increases as a result of ET. This is expected since the variable cost of the mid-merit and the peaker capacities clearly increases compared to the case when there is no ET. Another observation is that the relative share of peaker capacity increases compared to the mid-merit capacity as a result of ET. This is because the load for buying emission permits is higher for mid-merit than peaker technologies. We have assumed that mid-merit technology uses peat and coal as a fuel and fuel-based CO_2 multipliers for these are relatively higher than for the fuels on which peaker power technology is based.

Table 4.4 Effects of RTP with emission price 30 €/tCO_2

Share of RTP	Flat rate (€/MWh)	Total annual energy consumed (TWh)	Mid-merit power (MW)	Peaker power (MW)	Total equilibrium capacity (MW)	Peak price (€/MWh)	Hours at peak quantity (at 8760)
baseline	66.53	385.31	8,212	20,601	62,543	–	–
0.333	66.07	385.80	8,305	18,300	60,297	6580.39	53
0.666	66.18	386.07	8,356	16,961	58,978	2403.89	159
0.999	66.29	386.27	8,413	16,076	58,124	1258.46	285

Table 4.5 Effects of emission permit price, $\alpha = 0.333$

Emission permit price (€/tCO2)	Flat rate (€/MWh)	Total annual energy consumed (TWh)	Mid-merit power (MW)	Peaker power (MW)	Total equilibrium capacity (MW)	Peak price (€/MWh)	Hours at peak quantity (at 8760)
No ETS	43.09	390.08	14,419	12,799	60,753	6468.40	60
23	60.88	386.60	10,246	16,604	60,385	6545.63	55
30	66.07	385.80	8,305	18,300	60,297	6580.39	53
50	79.97	383.94	0	26,392	60,093	6657.61	48
100	109.85	380.92	0	26,037	59,753	6830.44	43

Table 4.6 Welfare effects of the RTP, no emission trade

Elasti-city	Share on RTP	Billing costs for RTP customers (million €)	Billing costs for flat rate customers, (million €)	Total billing costs (million €)	Producers' profits hydro (million €)	Producers' profits nuclear (million €)	Total producers' profits (million €)	Change in consumers' billing costs (from all in flat rate) (million €)	Change in producers' profits (from all in flat rate) (million €)
Baseline scenario	–	–	16988.89	16988.89	3673.20	1160.15	4833.35	–	–
−0.025	0.333	5531.11	11190.67	16721.79	3689.47	1168.79	4858.26	−267.10	24.91
−0.025	0.666	11100.27	5591.19	16691.46	3703.50	1176.25	4879.74	−297.42	46.39
−0.025	0.999	16660.83	16.77	16677.60	3717.72	1183.80	4901.52	−311.28	68.16
−0.1	0.333	5549.88	11284.99	16834.87	3754.06	1203.12	4957.18	−154.02	123.82
−0.1	0.666	11180.92	5648.27	16829.20	3801.64	1228.40	5030.04	−159.69	196.69
−0.1	0.999	16808.97	16.97	16825.93	3836.56	1246.95	5083.51	−162.95	250.16
−0.3	0.333	5661.80	11545.09	17206.89	3869.05	1264.21	5133.26	218.01	299.91
−0.3	0.666	11391.67	5774.64	17166.31	3916.57	1289.46	5206.03	177.43	372.68
−0.3	0.999	17138.99	17.36	17156.35	3960.34	1312.72	5273.05	167.46	439.70

Table 4.7 Welfare effects of the RTP with emission trade

Elasti-city	Share on RTP	Billing costs for RTP customers (million €)	Billing costs for flat rate customers, (million €)	Total billing costs (million €)	Producers' profits hydro (million €)	Producers' profits nuclear (million €)	Total producers' profits (million €)	Change in consumers' billing costs (from all in flat rate) (million €)	Change in producers' profits (from all in flat rate) (million €)
Baseline scenario	–	–	23637.98	23637.98	7018.97	2937.92	9956.88	–	–
−0.025	0.333	7776.76	15671.93	23448.69	7055.41	2957.28	10012.70	−189.29	55.81
−0.025	0.666	15606.88	7840.93	23447.81	7085.83	2973.45	10059.28	−190.18	102.40
−0.025	0.999	23439.00	23.54	23462.54	7117.06	2990.04	10107.09	−175.44	150.21
−0.1	0.333	7578.15	15325.48	22903.63	7039.56	2948.86	9988.42	−734.36	31.54
−0.1	0.666	15308.05	7703.76	23011.81	7164.10	3015.03	10179.13	−626.18	222.25
−0.1	0.999	23082.01	23.22	23105.23	7262.87	3067.51	10330.38	−532.76	373.50
−0.3	0.333	7168.48	14537.34	21705.82	7009.53	2932.90	9942.44	−1932.16	−14.45
−0.3	0.666	14572.84	7344.06	21916.90	7276.42	3074.71	10351.13	−1721.08	394.25
−0.3	0.999	21993.57	22.14	22015.71	7413.85	3147.74	10561.59	−1622.27	604.70

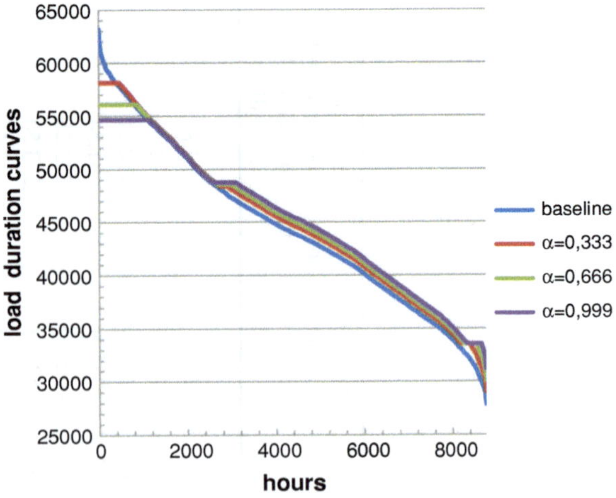

Fig. 4.2 Load duration with different share of the customers on the RTP, $\varepsilon = -0.1$, no emission trade

Fuel-based CO_2 multipliers are the following: for peat 382 $kgCO_2/MWh$, for coal 341 $kgCO_2/MWh$, and for oil 279 $kgCO_2/MWh$.

The results concerning total equilibrium capacity, peak price, and hours at peak quantity are rather similar to those of the RTP without ET. One more clear difference is that (keeping the RTP share constant) total quantity consumed starts to decrease with the elasticity value of −0.1. This result is strengthened when customers become more price sensitive. Interestingly, when the share of the customers on RTP increases, with the lower demand elasticities total demand increases, while the opposite is true for the demand elasticity of −0.3. Thus, if policy seeks to decrease the total amount of electricity consumed we should be careful when introducing RTP to a market where demand is very inelastic; it may increase the total demand. However, the amount of total capacity installed decreases as a result of RTP and this can improve the cost efficiency of investments. These results can be observed from Figs. 4.4 and 4.5. Especially from Fig. 4.5 one can clearly see the impact of demand elasticity to the load duration curve. With the elasticity value of −0.3 load curve is at almost all hours below the baseline.

An interesting question is whether RTP and Tradable Emission Pricing have different energy system efficiency effects or do they have any simultaneous effects. To see more clearly the impacts which result from RTP and ET, we have reported these simulation results separately in Tables 4.4 and 4.5. We have simulated the model with emission prices 23, 30, 50, and 100 €/tCO_2. The impact of RTP is shown by changing the share of customers on RTP while keeping the emission price constant at 30 €/tCO_2. We also keep the elasticity of demand in these simulations constant at −.025.[4] We want to compare RTP and ET here as two

[4] In the following we continue with the assumption that hydro- and nuclear power are constrained as in the previous simulations and they are not reported separately here.

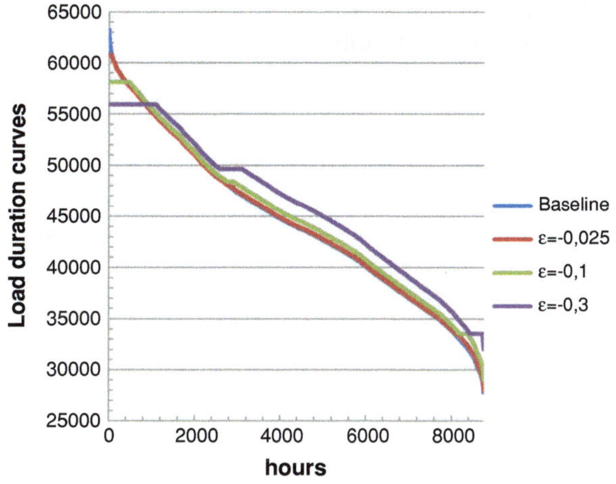

Fig. 4.3 Load duration with different values of elasticity, $\alpha = 0.333$, no emission trade

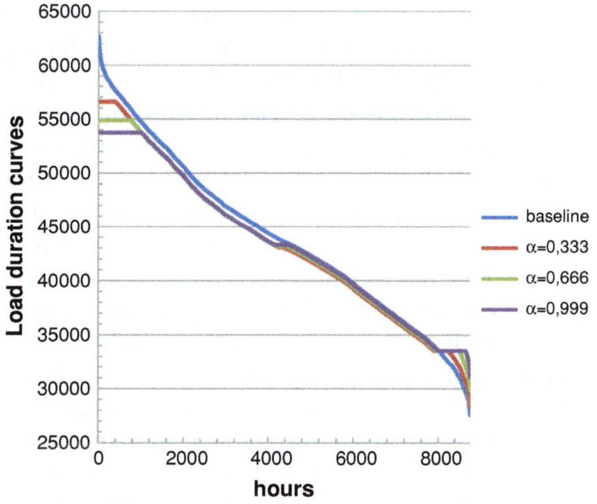

Fig. 4.4 Load duration curves with different share of the customers on the RTP, $\varepsilon = -0.1$, with emission trade

different policy instruments. From Table 4.4 we see that the RTP clearly reduces the need for peaker capacity and total equilibrium capacity. As the share of the customers on RTP increases the amount of total capacity as well as the amount of peaker capacity clearly diminishes. Another effect of RTP is to decrease the peak hour price and increase the number of hours at peak quantity. These effects can clearly be seen as the share of customer on RTP increases.

In Table 4.5 we report the simulation results with different emission permit prices. We keep here the share of RTP customers (33%) as well as elasticity of demand (−0.025) constant. As can be seen from Table 4.5 the impact of increases on emission price is more on the relative share of different types of capacities. As the emission price increases the relative share of peaker capacity increases and share

Fig. 4.5 Load duration with different values of elasticity, $\alpha = 0.333$, with emission trade

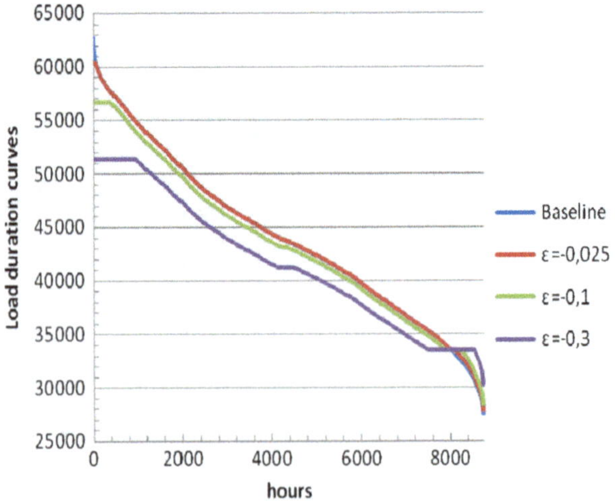

of mid-merit capacity decreases. This is because the representative mid-merit technology is (at least in our case) based on coal and peat which are more carbon-intensive fuels than the ones used in peaker technologies. Another clear observation to note is that the flat rate increases significantly as a consequence of increases in emission permit price. The increase of flat rate in turn reduces total annual consumption. The increase of emission permit price also reduces total capacity somewhat and increases the peaker hour price. These effects are, however, clearly smaller than the ones received by increasing the share of the customers on RTP.

Our results clearly show that RTP and ET have different effects on the energy system. RTP has its main effects on the peaker prices and hours while ET allocates energy demand away from mid-merit capacities. Thus these policy instruments should not be seen as two competing or substituting measures for emissions reductions but as complementary instruments. A simultaneous package of both has the best efficiency results from system efficiency and emission reduction efficiency aspects.

In Tables 4.6 and 4.7 we present welfare effects of RTP. First in the Table 4.6 we present the result based on RTP simulations without ET and then in the Table 4.7 the results with ET. From Table 4.6 we see that total billing costs of customers (compared to the baseline where all customers face a flat rate) decrease with the lower elasticity values and RTP but total billing cost for customers is higher with the elasticity value of -0.3. This is mainly due to the increase of total consumption and partly by the increase of the flat rate (compared to the lower elasticity value).[5] The billing cost for one MWh of the RTP customers, however, decreases as demand becomes more elastic.

[5] Our flat rate and welfare results behave somewhat differently than those of Borenstein [5] and Borenstein and Holland [6] because we have some constrained technologies whereas in their simulations all technologies are freely optimized. It is, however, not realistic in the Nordic power market context to freely optimize all technologies since considerable part of the production comes from hydro- and nuclear power stations.

Next in Table 4.6 we present the rents of the capacity constrained power units, i.e., units which are based on hydropower and nuclear power technologies. It can be seen that both technologies are making quite significant profits. Notable is that the profits are not very sensitive to the different demand elasticities and RTP scenarios. The variation of the profits mostly results from the change in the amount of peak hours. The more elastic the demand and the more customers on RTP, the larger is the amount of peak hours. These are the hours when the market clearing price is above marginal costs of these technology types and consequently from those hours it is possible for the technologies which are lower on the supply function to make profits.

Last in Table 4.6 we present the changes in consumers' billing costs and producers' profits from the case when all customers face a flat rate. The change in consumers' billing costs is negative except with the elasticity value of −0.3. This is, however, as already mentioned above, mainly due to the increase in total consumption and not as a result of increase in per MWh price of the RTP customers. Change in the producers' profits is positive and increasing for all elasticity and RTP values. Also the profits per installed MW increase as demand becomes more elastic and more customers belong to the RTP group.

In Table 4.7 we present welfare results of the RTP simulations with ET. When we compare the welfare results of the RTP with and without ET we see some clear differences. The first difference is that now the annual total billing cost of customers' decreases with all elasticity and RTP values. This is mainly due to the decrease in total consumption (starting from the elasticity of −0.1) and partly due to the decrease in per MWh price for RTP customers. The second difference is the clear increase in the profits of both hydro- and nuclear power. The increase of profits is strengthening with combination of higher elasticity value and higher share of the customers in RTP. If we compare the total profits of baselines with and without ET, we see that profits increase by 5,123 million€. This means that hydro- and nuclear power producers make quite significant windfall profits and clearly benefit from the ET. As a summary we can say that RTP with ET has two different kinds of effects which may have politically opposite signs. On the one hand, policy goals targeting energy conservation require customers with higher price elasticity, but at the same time if price elasticity increases we see higher windfall profits.

References

1. Allcott H (2011) Rethinking real-time electricity pricing. Resour Energy Econ 33:820–842
2. Borenstein S (2009) Electricity pricing that reflect its real-time cost. NBER Reporter, Winter, 9–12
3. Borenstein S (2007) Customer risk from real-time retail electricity pricing: bill volatility and hedgability. Energy J 28:111–130
4. Borenstein S (2007) Wealth transfers among large customers from implementing real-time retail electricity pricing. Energy J 28:131–149
5. Borenstein S (2005) The long-run efficiency of real-time electricity pricing. Energy J 26:1–24

6. Borenstein S, Holland S (2005) On the efficiency of competitive electricity markets with time-invariant retail prices. Rand J Econ 36:469–493
7. Borenstein S, Bushnell J (1999) An empirical analysis of the potential for market power in a deregulated california electricity industry. J Ind Econ 47:285–323
8. Borenstein S, Bushnell J, Wolak F (2002) Measuring market inefficiencies in california's deregulated wholesale electricity market. Am Econ Rev 92:1376–1405
9. Eurelectric (2006) Latest industry statistics as at 31 December 2006. available at http://www.eurelectric.org/Statistics/Latest2006.htm, Accessed 29 Aug 2011
10. Faruqui A, Hledik R, Tsoukalis J (2009) The power of dynamic pricing. Electricity J 22:42–56
11. Herriges JA, Baladi SM, Caves DW, Neenan BF (1993) The response of industrial customers to electric rates based upon dynamic marginal costs. Rev Econ Stat 75:446–454
12. Holland S, Mansur ET (2008) Is real-time pricing green? The environmental impacts of electricity demand variance. Rev Econ Stat 90:550–561
13. Joskow P, Kahn E (2002) A quantitative analysis of pricing behavior in california's wholesale electricity market during summer 2000. Energy J 23:1–35
14. Kopsakangas-Savolainen M, Svento R (2011) Real-time pricing in the nordic power markets. Energy Econ. doi:10.1016/j.eneco.2011.10.006
15. Nord pool (2011). www.nordpool.com, Accessed 15 May 2011
16. Nord pool spot (2011). www.nordpoolspot.com/reports/consumption/, Accessed 15th May 2011
17. Patrick RH, Wolak F (1999) Consumer response to real-time prices in the england and wales electricity market: implications for demand-side bidding and pricing options design under competition. In: Crew MA (ed) Regulation under increasing competition. Kluwer Academic Publishers, Dordrecht
18. Schwarz PM, Taylor TN, Birmingham M, Dardan SL (2002) Industrial response to electricity real-time prices: short run and long run. Econ Inq 40:597–610
19. Tarjanne R, Kivistö A (2008) Comparison of electricity generation costs, Research report EN A-56. Lappeenranta University of Technology, Finland

Chapter 5
The Effects of Nuclear Power Investments in Real-Time Pricing Framework

5.1 Background

Investments in conventional energy technologies are bound to involve questions of hazardous and even catastrophic outcomes. Nuclear catastrophes and climate change are the evident examples. Economics of hazardous outcomes has advanced in the past decades (e.g., [3, 4, 9, 11]). It is fair to say, however, that the economic research is still far away from classroom solutions, but at the same time it is quite clear that the right questions have been identified. These relate to preference descriptions, discounting, loss functions, and non-familiar probability distribution assumptions.

The new wave of discounting research and discussion started from the Stern Review [7]. The review concluded from analysis based on a large computer simulation model PAGE that "… the overall costs and risks of climate change will be equivalent to losing at least 5% of global GDP each year, now and forever. If a wider range of risks and impacts is taken into account, the estimates of damage could rise to 20% or more." (p. xv).[1] This is in sharp contrast with conventional "Integrated Assessment Models" the most well-known of which is the DICE model of Nordhaus [5, 6]. The DICE solution can be described by what has become to be known as a "policy ramp" option by gradually tightening emissions over time. The basic argument of this solution is that by investing in current technologies the world can, at the same time, increase growth and learn more about the possibilities to solve global warming problems in future.

An interesting feature of these models strongly pointed out by Dasgupta [3] is that even though they are both large computer models the core ethics of inter-generational welfare economics that these models use can be characterized by only two numbers: the social time discount rate and the elasticity of marginal felicity (utility) of consumption. Assuming the elasticity of marginal felicity to be constant

---

[1] Dasgupta [2, 3] shows how interestingly this analysis is very close to that proposed by [1].

M. Kopsakangas-Savolainen and R. Svento, *Modern Energy Markets*,
Green Energy and Technology, DOI: 10.1007/978-1-4471-2972-1_5,

and maximizing intergenerational welfare it can be shown that the underlying structure of welfare is then additive in the sense that the equilibrium rate of return on investments (and discount rate of consumption) is a sum of the social discount rate and by growth of consumption-weighted elasticity of marginal utility of consumption.

Stern Review and Norhaus both use value one for the elasticity but very different values for the social time preference: Stern 0.1% a year and Nordhaus 3% a year. Combining these to their standard growth assumptions we have the consumption discount rates in these models as 1.4% a year for Stern and 4.3% for Nordhaus. An example related to these numbers is given by Dasgupta [3]. A given loss in consumption realizing after 100 years from now and discounted by 4.3 is 17 times smaller than the same loss discounted by 1.4. Since both models use identical marginal utility elasticities their differences in results and policy recommendations degenerate to their differences in social time discount rates.

The low social discount rate used by Stern can be defended most strongly on ethical and philosophical grounds. It can be argued that social impatience is ethically indefensible. To choose the social discount rate as non-negative is the same as to favor policies that discriminate against the well-being of future generations on the simple ground that they are not present today. This argument of course bypasses the questions of economic growth and technical progress which should work in the direction of making the future generations richer than the current one. Nordhaus bases his choices much more on real economy facts. Nordhaus insists [5, 6] that these parameters of social discount rate and elasticity of marginal utility must be chosen to be consistent with: (1) market interest rates, (2) observed values of consumption growth, and (3) rates of private and public saving and investment. Thus Nordhaus's modeling logic rests strongly on observed market-based facts. This is of course sound economics in the sense that the market outcomes by definition are based on revealed preferences and thus give a solid ground for analysis. But other objections can of course be raised. First of all, climate change is a common problem caused by activities and investments that create the externalities generating this tragedy of the commons in the first place. And as we know, markets do not price these externalities correctly. Backward-looking analysis also escapes the question of time-varying preferences. This preference question is, however, equally unsolved in economic long-run analyses in a more general sense as well.

There is also a line of research analyzing the possibility of time varying or hyperbolic development of the discount rate itself [8, 10]. Based on opinions of economists Weitzman [8] concluded that the probability density function for an uncertain discount rate would be a gamma distribution and this gives the declining discount rate schedule as a simple closed-form solution of time. In his paper [10] Weitzman extends this approach by using a Ramsay optimal growth model. He shows that risk-adjusted gamma discounting on lowering distant future discount rates may be significant and that the driving force is a "fear factor" from risk aversion to catastrophic future states of the world. The basic message is that the distant future should be weighted more heavily than is done by standard exponential discounting at a constant rate.

Weitzman [9, 11] also opens the question of the role of structural uncertainties evidently related to climate change modeling. By this he means that there are many unknown unknowns and that we have to use probability distributions over probability distributions in the analysis. Various mechanisms related to these uncertainties will produce a fat-tailed distribution of long-run temperature changes. Weitzman emphasizes [9, p. 12] that the critical question is *how fast* does the probability of a catastrophe decline relative to the scope and impact of it. A thin-tailed probability density function is of less concern because the probability of the bad event declines exponentially (or faster). Should the probability decline polynomially producing a fat tail can be a much more worrisome state of affairs for policy recommendations. As Weitzman himself puts it:

> Compared with the thin-tailed case cost-benefit analysis of fat-tailed potential catastrophes is inclined to favor paying a lot more attention to learning how fat the bad tail might be and—if the tail is discovered to be too heavy for comfort after the learning process—is a lot more open to at least considering undertaking serious mitigation measures (including, perhaps, geoengineering in the case of climate change) to slim it down fast. The key economic questions are: what is the overall cost of such a tail-slimming weight-loss program and how much of the bad fat does it remove from the overweight tail? (ibid. p. 24)

Dasgupta [3] has also analyzed future uncertainties related to intergenerational well-being. His results once again remind us of the "dismal science" features of economics. Assuming that the uncertainty is not related to realizations over constant consumption growth paths, but on productivity and output, Dasgupta shows that neither stochastic nor risk-free consumption discount rates along optimum consumption sequences are hyperbolic. He shows that in an uncertain world the elasticity of marginal utility of consumption plays a double role. It is not only an index of inequality aversion but it is also an index of risk aversion. Larger values of this elasticity recommend earlier generations to save less for the future (the equity motive), whereas larger values of it recommend earlier generations to save more (the precautionary motive) (ibid, Proposition 3). Dasgupta furthermore shows that even a thin-tailed distribution for the discount rate combined with "large" risk produces the result that no optimum policy exists if the variance of the discount distribution is bounded from below. This means that the problem of optimum saving, when formulated in terms of expected well-being over an infinite horizon, is so inadequately posed as to defy an answer. Consumption discount rates cannot be defined and social cost benefit analysis of projects becomes meaningless.

Needless to say both models used by Dasgupta and Weitzman are openly simplistic and these results cannot be taken as the whole truth and both researchers of course are very well aware of this. But they are important and influential in showing probable roads for future research. At least three such lines can be identified. The first one relates to analyzing the policy conclusions more openly related to inequality questions in the sense that there must be a non-contradictory solution for both intra-generational and inter-generational equalities. This also relates to new openings concerning modeling of the preferences in the sense of allowing the elasticity of marginal utility of consumption to be dynamic in the sense of being time- or consumption level-dependent, or both. The third line of research relates to identifying different kinds of plan (B)s in the case of really bad realizations for the state of the world.

One line of research related to these difficult questions that in any case can be taken, and the one we shall pursue here, relates to the efficiencies of the possible policies related to climate change. Not much research has been devoted to the question of simultaneous effects of various policies. This is especially true when it comes to policies related to modern technologies and current possibilities of using them. In this chapter we enlarge the basic Real-Time Pricing energy production model presented in Chaps. 3 and 4 by relaxing the earlier assumption concerning nuclear power capacity limits. Here, we allow new nuclear power investments and analyze their effects to the capacities, prices, and emissions of the Nordic power market in combination with Real-Time Pricing of electricity. We are in an interesting situation here in the sense that we have a real case to analyze. In 2010, the Finnish parliament decided to allow construction of two new nuclear reactors. Taking this decision the Finnish political system revealed that nuclear catastrophes are either discounted with a higher rate and/or the probability distribution has a thinner tail compared to climate change. At the same time, of course, Finland is a small open economy with important energy-dependent export industries.

5.2 Nuclear Power Scenarios

There are five nuclear power plants currently operating in Nordic countries. Three of them (10 reactors) are located in Sweden, whereas two of them (4 reactors) are located in Finland. There is also one more 1,600 MW reactor under construction with permission to construct two more reactors in the near future in Finland. Simulations in this section are based on two different scenarios (scenarios 2 and 3) regarding nuclear power production. The results of these scenarios are compared to the results derived from the current nuclear power structure (presented in Table 4.3).

The current nuclear power structure is referred as scenario 1 in what follows. The first new scenario (referred as scenario 2) is based on the assumption that nuclear power is constrained to the level after the nuclear power reactor under construction (Olkiluoto 3) and the two already granted new construction permits (TVO and Fennovoima) are realized in Finland. The second new scenario (referred as scenario 3) for nuclear power capacity is based on the calculation of how much nuclear capacity must be built if we want to replace conventional mid-merit power plants completely by nuclear power. The cost data and load profile used in the simulations are the same as used and presented in Chap. 4 (see Table 4.1 and Fig. 4.1).

5.3 Empirical Results

In Table 5.1 we first present the baseline values which express the situation when all customers are facing a flat rate. The capacity values presented in the first line hence present capacities needed to efficiently provide electricity for the demand under flat rate and assuming the cost structure presented in Table 4.1. The next lines present the equilibrium prices, capacities, and other information under nuclear scenarios 2 and 3 with different assumptions of price elasticity of demand and the proportion of customers on RTP. We assume, as in simulations in Chap. 4, that the hydro capacity is constrained to the level of 47,816 MW of which on average 21,899 MW is in use. The capacity of nuclear power has been increased from 11,636 MW (capacity in the simulations in Chap. 4) to 16,676 MW in scenario 2 and then to the amount required to replace mid-merit technology completely in scenario 3. For emission calculations we have assumed that mid-merit technology uses peat and coal as a fuel and fuel-based CO_2 multipliers for these are relatively higher than for the fuels on which peaker power technology is based. Fuel-based CO_2 multipliers are the following: for peat 382 kgCO_2/MWh, for coal 341 kgCO_2/MWh, and for oil 279 kgCO_2/MWh.

The clear difference between the results of scenarios 1 and 2 (see Tables 4.3 and 5.1) is that when more nuclear capacity is built the flat rate decreases systematically. Another result is that the amount of mid-merit capacity is clearly smaller when some new nuclear capacity is installed. This does not, however, seem to have much impact on the need for peaker capacity. This is understandable since so far as the amount of nuclear capacity is reasonable it does not affect the shape of the peak demand loads. From Table 5.1, scenario 2, we can see that again the more elastic the demand is and the more customers are on RTP the smaller is the efficiency amount of the total capacity. Also, the need for peaker capacity is smaller the more elastic the demand is and the more customers belong to the RTP group. Also, the peak price and hours at peak demand behave similarly, with respect to different demand elasticity and RTP values, to those of the basic nuclear scenario. Notably, when some new nuclear capacity is installed the total electricity demand increases compared to nuclear scenario 1. This might be mainly due to two reasons. First the real-time price of lower demand hours clearly decreases as some mid-merit technology with higher variable costs is replaced by lower variable costs nuclear capacity and consequently the total demand increases. Another explanation comes from the systematic decrease in flat rate and as a result the flat rate customers also increase their consumption. If we compare the emissions resulting from baseline scenario, where all customers face flat rate and nuclear capacity is at the current level of 11,636 MW we see that when the amount of nuclear capacity increases to 16,676 MW CO_2 emissions decrease significantly. Interesting is that the decrease in emission is strengthened as the demand becomes more elastic and also with the share of the customers in RTP. This confirms the argument of importance of Real-Time Pricing and price-sensitive demand in reaching the EU-set targets of mitigation in greenhouse gases.

Table 5.1 Effects of nuclear capacity investments (scenarios 2 and 3)

Elasticity	Share on RTP	Total annual energy consumed (TWh)	Flat rate €/MWh	Nuclear power (MW)	Mid-merit power (MW)	Peaker power	Total equilibrium capacity (MW)	Peak price €/MWh	Hours at peak quantity (of 8,760 h)	Total emissions, million tCO_2
Baseline scenario	–	386.11	61.32	11,636	10,206	18,932	62,673	–	–	31.28
Nuclear scen.2										
−0.025	0.333	387.80	55.42	16,676	5,285	16,622	60,483	6,562.87	55	18.48
−0.025	0.666	388.36	55.56	16,676	5,282	15,224	59,081	2,374.79	164	18.41
−0.025	0.999	388.82	55.71	16,676	5,266	14,317	58,158	1,229.56	298	18.32
−0.1	0.333	385.12	55.37	16,676	4,754	13,622	56,952	973.32	387	17.02
−0.1	0.666	386.10	56.07	16,676	4,719	11,750	55,044	357.01	762	16.59
−0.1	0.999	386.50	56.81	16,676	4,725	10,440	53,740	235.85	1,003	16.25
−0.3	0.333	377.25	55.59	16,676	3,366	10,390	52,331	287.80	942	13.15
−0.3	0.666	374.76	57.45	16,676	3,214	7,546	49,335	154.78	1,568	11.71
−0.3	0.999	371.77	58.81	16,676	3,331	5,499	47,405	122.95	2,008	10.89
Nuclear scen.3										
−0.025	0.333	389.73	47.25	22,091	0	16,657	60,647	6,596.15	55	8.14
−0.025	0.666	390.58	47.66	22,006	0	15,255	59,160	2,378.33	163	8.07
−0.025	0.999	391.32	48.02	21,945	0	14,314	58,158	1,229.76	298	7.97
−0.1	0.333	391.69	48.41	21,822	0	13,766	57,487	988.31	382	7.87
−0.1	0.666	393.28	49.70	21,574	0	11,805	55,278	358.79	759	7.41
−0.1	0.999	394.40	50.81	21,400	0	10,442	53,741	235.84	1,001	6.96
−0.3	0.333	390.33	51.02	20,770	0	10,633	53,302	293.69	925	6.95
−0.3	0.666	387.01	53.82	20,164	0	7,626	49,689	155.45	1,561	5.60
−0.3	0.999	382.58	55.81	20,007	0	5,500	47,406	122.95	2,008	4.34

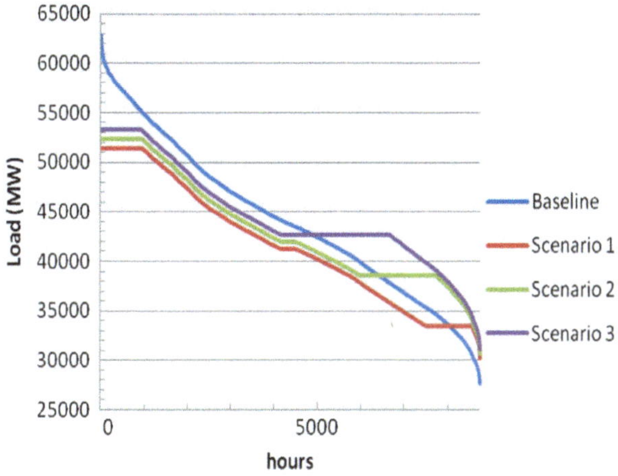

Fig. 5.1 Load duration curves, demand elasticity −0.3, $\alpha = 0.333$

The last nuclear scenario assumption is that the nuclear power units are built up to the point where they replace the former mid-merit power capacity totally.[2] The results for the simulations are presented in Table 5.1, nuclear scenario 3. The amount of nuclear power capacity needed to replace the former mid-merit capacity depends especially on the price elasticity of demand. The more elastic the demand the less we need nuclear capacity to replace former mid-merit capacity. Also the share of the customers in RTP has some impact on the need for nuclear power capacity. The need for nuclear power capacity to replace mid-merit capacity is a slightly smaller if relatively more customers belong to the RTP group. Interestingly, the needed quantity for replacement of mid-merit technology by nuclear power is about the same size (10 new reactors) as that to which some radical nuclear power advocates in Sweden have referred.[3] The simulation results concerning the amount of peak power capacity, peak price, and hours at peak quantity are rather similar to those in the former simulations presented in Tables 4.3 and 5.1. The clear difference is the level of the flat rate which is significantly lower based on the simulations of nuclear scenario 3. The CO_2 emissions are systematically lower under nuclear scenario 3 than under the other scenarios. Also here, the increase in price elasticity and share of the customers in RTP strengthen the decrease in emissions.

From Fig. 5.1 the impact of different nuclear power scenarios on the load duration curve can be observed. The most striking thing resulting from installation of new nuclear power is the increase in the total demand. This can most clearly be seen from the nuclear scenario 3 load curves. According to this load curve, demand increases especially in lower demand hours resulting in higher total annual demand. This is of course against the openly agreed target of energy conservation.

[2] This is a somewhat unrealistic assumption since especially in Finland most mid-merit-based electricity production is based on combined heat and power production.

[3] See e.g. www.world-nuclear.org/info for the discussion of the nuclear power role in Sweden.

Table 5.2 Profits of hydropower and nuclear power units under different nuclear scenarios and representative values of elasticity and share of RTP

Elasticity	Share on RTP	Hydro power, total profits in million € (total capacity 47,816 MW, average capacity on use 21,899 MW)	Hydro power, profits/ MW in €	Nuclear power, total profits in million €	Nuclear power, profits/ MW in €
Baseline scenario		7,018.97	146,791	2,937.92	252,485
Nuclear scenario 1					
−0.1	0.333	7,039.56	147,222	2,948.86	253,426
−0.1	0.999	7,262.87	151,892	3,067.51	263,623
−0.3	0.333	7,009.53	146,594	2,932.90	252,054
−0.3	0.999	7,413.85	155,050	3,147.74	270,517
Nuclear scenario 2					
−0.1	0.333	5,799.45	121,287	3,281.79	196,797
−0.1	0.999	6,172.17	129,082	3,565.61	213,817
−0.3	0.333	5,892.47	123,232	3,352.62	201,044
−0.3	0.999	6,109.67	140,350	3,975.90	238,421
Nuclear scenario 3					
−0.1	0.333	4,355.49	91,088	2,855.61	130,859
−0.1	0.999	4,880.42	102,067	3,313.36	154,830
−0.3	0.333	4,919.45	102,883	3,252.84	156,612
−0.3	0.999	6,028.66	126,080	4,146.72	207,264

In Table 5.2 we present the rents of the capacity constrained power units, i.e., units which are based on hydropower and nuclear power technologies. It can be seen that both technologies are making quite significant profits. Notably, the profits are not very sensitive to the different demand elasticities and RTP scenarios. The variation of the profits mostly results from the change in the amount of peak hours. The more elastic the demand and the more the customers on RTP, the larger is the amount of peak hours. These are the hours when the market clearing price is above marginal costs of different technology types, and consequently from those hours it is possible for the technologies which are lower on the supply function to make profits. When we compare the results under nuclear scenarios 1 and 2, one clear observation is that profit per MW decreases for both hydro- and nuclear power units. The percentage decrease is greater for nuclear power producers than for hydropower producers. However, again the profits for hydro- and nuclear power producers are higher the more elastic the demand and the more the customers in the RTP group.

Finally, in Table 5.2 we present the rents of the hydropower and nuclear power units under nuclear scenario 3 and with different demand elasticity and RTP values. The most significant observation is that the profit/MW of installed nuclear power decreases considerably especially with lower demand and lower share on the RTP. For example, with an elasticity value of −0.1 and with 0.333 share of the RTP customers, the decrease in profit/MW compared to scenario 2 is about 34%.

Notably, also profits for hydropower generators decrease as a result of the increase in nuclear power capacity. This of course reduces the need for possible introduction of windfall taxes. The percentage decrease in nuclear power profits is the bigger the less elastic is the demand. It seems that the share of the customers in RTP has increasing impact on the profits of baseload capacity.

In our simulations the units using capacity constrained technologies are able to make considerable profits. According to our results these profits are not very sensitive to the different demand elasticities or RTP scenarios. The most significant observation is that the profit/MW of installed nuclear power decreases when some new nuclear capacity is built. Also, profits for hydropower generators decrease as a result of the increase in nuclear power capacity. This should reduce the need for recently discussed windfall taxes in the Nordic power market.

Similar to some other studies we find that though RTP decreases the demand during peak hours this does not mean that the total consumption decreases. Actually, according to our results the total consumption increases with the share of the customers in RTP if demand is very inelastic. The total consumption, however, decreases with demand elasticity. New nuclear power investments have a clear role in changing total demand. We find that investment in new nuclear capacity increases the total demand somewhat. This is contrary to the public policy objective of energy conservation. Safety costs of nuclear power investments are difficult to estimate. In final policy analysis these should of course be included.

References

1. Cline WR (1992) The economics of global warming. Institute for International Economics, Washington
2. Dasgupta P (2007) Commentary: the Stern review's economics of climate change. Nat Inst Econ Rev 199:4–7
3. Dasgupta P (2008) Discounting climate change. J Risk Uncertainty 37:141–169
4. Environmental and Resource Economics (2009) Special issue 43
5. Nordhaus WD (1994) Managing the global commons: the economics of climate change. MIT Press, Cambridge
6. Nordhaus WD (2007) A review of the Stern review on the economics of climate change. J Econ Lit XLV:686–702
7. Stern N (2007) The economics of climate change: the Stern review. Cambridge University Press, Cambridge
8. Weitzman ML (2001) Gamma discounting. Am Econ Rev 91:260–271
9. Weitzman ML (2009) On modeling and interpreting the economics of catastrophic climate change. Rev Econ Stat 91:1–19
10. Weitzman ML (2010) Risk-adjusted gamma discounting. J Environ Econ Manage 60:1–13
11. Weitzman ML (2009) Some basic economics of climate change. In: Touffut J-P (ed) Changing climate, changing economy. Edward Elgar, UK

Chapter 6
Emission Trading and Market Access of Renewables

6.1 Background

In the previous chapter we studied how Real-Time Pricing (RTP) together with increased investments in nuclear capacity effect energy generation and emissions. Here we shall ask similar question but this time we shall look at the effects of RTP together with an emission trading scheme. We are especially interested on the common effect of these instruments on wind power emerging on the market.

The methodology is based on the RTP energy production simulation model presented above in Chap. 3 and the application is done by using cost and load profile data (presented in Chap. 4) from Nordic power markets. The model differs from the basic model in the sense that in this chapter we include a volatile energy price source (wind power) into our model. We study the effects of emission trading in promoting market access for wind power. The model and algorithm used is similar to those used in previous Chaps. 4 and 5. The only difference is the possibility for new technology, wind power, to enter the market.

As Perdan and Azapagic [16] note there are currently five mandatory emissions trading systems (ETSs) operating in the world. These are: The European Union ETS [4–7], the Regional Greenhouse Gas Initiative in the USA [19–22], New Zealand Emissions Trading Scheme [14, 15], Tokyo metropolitan trading scheme [25–27], and the New South Wales Greenhouse Gas Abatement Scheme [12, 13]. These schemes naturally differ in various ways; they have different sectoral and temporal coverage and different emissions targets. They do have one common premise, though, they all aim at lowering the overall cost of combating climate change by ensuring that emission reductions take place where the cost of reduction is the lowest.

M. Kopsakangas-Savolainen and R. Svento, *Modern Energy Markets*,
Green Energy and Technology, DOI: 10.1007/978-1-4471-2972-1_6,

6.2 On Emission Trading Schemes in Europe

Convery [2] gives a comprehensive look at the origins and development of the EU ETS. The development and birth of the current emissions trading schemes is interesting and in some senses almost paradoxical. During a quite narrow time span emissions trading evolved in EU from a non-considered policy option to the cornerstone of climate change policy. At the same time US refuses to ratify international treaties and the original proponent of trading schemes has turned largely to voluntary measures in its portfolio of climate change policies. The Kyoto Protocol can be seen as the key driver behind these diverging routes.

The European Commission was an early proponent of quantitative restrictions on greenhouse gas emissions. In 1992, the commission proposed an EU-wide carbon energy tax and argued at the Rio summit in the same year that quantitative restrictions should be added to climate change management policy instruments. But it soon became obvious that member states would not allow taxing rights to the union fearing that this would open up the road for other fiscal policy harmonizations. At the same time many nation states started to plan their own emission trading schemes. The Kyoto Protocol then provided the necessary instrument for the EU to take its leading role back. The Kyoto Protocol was directly usable in the sense that it provided a quantitative target from 1990 emissions for EU 15, flexible mechanisms including emissions trading and an impetus for the burden sharing mechanism.

The EU ETS process moved to center stage after US rejected the Kyoto Protocol in 2001. Since this rejection implied that the Protocol had to be accepted by all major players, especially Japan, Canada, and Russia, EU took a central and leading role in the following negotiations. In this process the European Union showed that it could, at the time, play an innovative and effective role in reaching solutions to international challenges.

In March 2000, the Commission launched a Green Paper on emissions trading and in October 2003, the Directive 2003/87/EC on the European Parliament and the Council establishing a scheme for greenhouse gas emissions trading within the Community came into effect.[1] The EU ETS became operational on Jan 1, 2005 and it is being implemented in three distinct phases. The first "pilot" phase covered the years 2005, 2006, and 2007. The primary goal was to develop the infrastructure and provide the experience for the later more serious phases. About 11,000 installations in 25 EU member states were covered. The allowances were given for free and their allocation to individual installations was left for the member states to do.

The second phase covered the years 2008–2012 and during this period the volume of emissions allowances was cut to 6.5% below the 2005 level. The scope of the scheme remained unchanged, with the exception of emissions of nitrous oxide from the production of nitric acid which are now included.

[1] Green Paper on greenhouse gas emissions trading within the European Union, COM (2000) 87 Final.

Phase three will run for eight years up to 2020. The scheme shall be strengthened and extended enabling it to be the central instrument in achieving the climate and energy targets set for 2020. More industrial sectors[2] and greenhouse gases shall be added. The Aviation Directive [6] adds the aviation industry into the EU ETS. The big difference, however, is the change to an auction mechanism as the allocation instrument.

Why did European Union succeed in implementing the EU ETS? Convery [2, pp. 407 and 408] gives what he thinks "seem to be amongst the key explanations:

- Free allowances—this met the needs of most industrial emitters.
- Fear of the alternative—the counterfactual. The use of carbon taxes and/or regulation (command and control) was variously proposed as alternative means of reducing emissions. Either would be less attractive to much of industry than trading.
- Information flow from the US generally and US business in particular, based on the acid rain experience.
- Many Member States, and especially the smaller ones, tend to support the Commission unless there are major strategic reasons to do otherwise.
- EU ETS was conflated with the Kyoto Protocol in some rhetoric and many minds, so that to support Kyoto was synonymous with support for emissions trading.
- The use of an obligatory three year Pilot Phase provided a real test of the EU ETS in action, a mechanism for temporarily indulging some Member State preoccupations including opt out and pooling, and a way of identifying weaknesses and correcting for some.
- … a surprising lack of overt attention to the fact that, in some jurisdictions, the opportunity cost of free allowances was likely to be passed through in part in the form of higher electricity prices to consumers.
- The ability to make tradeoffs on the world stage—and in particular to secure Russian agreement to ratify the Kyoto Protocol in exchange for European support for World Trade Organization (WTO) membership." (for the complete list see Convery [2]).

Measured by value and volume EU ETS has become the largest carbon market in the world. Already in the first year 2005, at least 362 million allowances were traded with the value of 7.2 billion Euros. The trading volume rose to 1 billion allowances in 2006, to 1.6 billion in 2007, over 3 billion in 2008 and to 5.6 billion in 2009 [17]. Concerning the emissions data points to a 2–5% decline in emissions (40–100 Mt CO_2 annually) [11].

The question of the link between electricity prices and (opportunity cost) carbon prices has been widely studied [1, 8, 18, 23, 24]. The results are, however, mixed and the main explanation can be sought from coexistence of several

[2] Dijkstra et al. [3] have done an interesting analysis related to sectoral broadening of the scheme.

electricity markets in Europe and the heterogeneity of national energy mixes. Kirat and Ahamada [9] have studied the pass-through question in the electricity markets of Germany and France. They show that the impact during the pilot phase was dependent on the country's energy mix. They show how this impact was felt in two phases the first of which covers years 2005–2006 and when the cost of carbon was included in the generation cost functions. In the second phase after 2006 this effect did not show up. Generators using more fossil fuels were more likely to include the price of emission permits in their electricity generation cost functions. Problems related to implementation of EU ETS—excess allocations and impossibility of "banking"—contributed to the collapse of the carbon spot price at the end of the pilot phase relaxing the carbon coercion to which generators were subject to in the beginning of implementation of EU ETS [9, p. 1003].

6.3 RTP, Emissions Trading, and Wind Power Market Access

Nordic power markets naturally operate under European Commission's internal emissions trading. At present, the emissions trading in Nord Pool only concerns the carbon dioxide emissions. The emissions trading scheme is meant to operate so that the emissions of the companies under the scheme keep the predefined total emissions quantity within the limits. For electricity markets the Emissions Trading Act is applied to carbon dioxide emissions of such power stations for which thermal input is more than 20 MW and also for the smaller combustion installations connected to the same district heating network. Typically the issuance of permits lies with the national Energy Market Authority. The amount of issued permits by power stations is less than their yearly emissions. Power producers can buy extra permits from the emission permit markets. This increases the costs of technologies under emission trade. Consequently, emission trade affects the profitability of wind power through making it more profitable for wind power producers to produce on those hours when conventional thermal power is the market clearing technology as can be seen from Fig. 6.1.

For wind power, we have assumed that specific investment costs are 1,400 €/kW. For a 20 year economic lifetime and annual real interest rate of 3% we end up with annual capital costs of 94,102 € per installed MW. For operation and maintenance costs we have assumed 4 €/MWh. There are some clear differences between modeling wind power and conventional thermal power producers profit maximization. The first difference is that wind power can be produced only when the wind is blowing. We assume that the utilization time of wind power per year is 2,500 h. In the simulations we have assumed that wind power is allocated to the first 2,500 h out of the 8,760 h in the load duration curve. This means that wind power in our simulations is used in those hours when demand is smallest. This can be justified as follows: this kind of procedure gives a lower bound to the profits (revenue) of wind power producers and in the case that wind power supply does not happen in these first 2,500 h, there will always be enough other technologies to

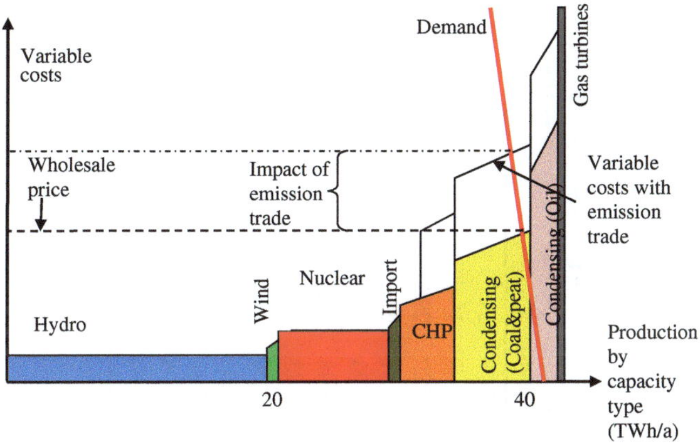

Fig. 6.1 Nordic electricity markets with emission trade

replace wind power. This is because the total amount of capacity has to be built in order to satisfy demand in highest hours and in low demand hours most of this capacity is unutilized. In order to justify this wind power production in the first 2,500 h all we need is to assume that it will wind at least 2,500 h during the year. Wind power producers cannot do worse if the wind is allocated to later hours in the load curve. Relying on these arguments it is safe to assume that wind is allocated to the first 2,500 h in the load curve.

We have looked at the impact of different carbon emission prices combined with the RTP to the promotion of renewable energy (wind in our case) to enter the Nordic power market. The aggregate target level for wind power capacity in Denmark, Finland, Norway, and Sweden is 25,600 MW by 2020. We have simulated the model with the carbon prices 23, 30, 50, and 100 €/tCO_2. The simulation results with different emissions prices and wind power technology are presented in Table 6.1 and in Fig. 6.2. We have assumed that demand is inelastic with the elasticity value of −0.025.

The first thing to notice is that regardless of the carbon price the amount of wind power entering to the market increases with the share of the consumers on RTP. The second finding is that the amount of wind power entering to the market is really increasing with the emission permit price. We can thus conclude that introducing RTP is not thus just important on the efficiency of investments of the peaker capacity but also as a method to promote renewable energy to the market if it is combined with the sufficiently high carbon price. On the capacity structure point of view, as the carbon price increases to 50 €/tCO_2 it is not profitable for the mid-merit power to enter the market. This is because midmerit is based on coal and peat which are more carbon intensive than peaker power and thus the impact of increase in carbon price increases their costs significantly. The increase in carbon price does not have significant impact of total equilibrium capacity required neither to the peak price or hours at peak quantity. These are more sensitive to the share of the customers on RTP.

Table 6.1 Emission permit price, RTP, and wind power

Emission permit price (€/tCO_2)	Share on RTP	Flat rate (€/MWh)	Total annual energy consumed (TWh)	Wind power (MW)	Mid-merit power (MW)	Peaker power (MW)	Total equilibrium capacity (MW)	Peak price (€/MWh)	Hours at peak quantity (at 8,760)	Total emissions (million tCO_2)
23	0.333	60.76	386.62	0	10,093	16,605	60,387	6545.78	55	30.81
23	0.666	60.79	386.93	442	10,121	15,211	59,035	2371.98	164	30.82
23	0.999	60.73	387.18	1,874	9,976	14,306	58,157	1230.29	298	30.35
30	0.333	65.36	385.94	3,669	7,607	18,291	60,308	6584.81	53	27.92
30	0.666	65.28	386.30	4,397	7,416	16,964	58,985	2404.23	159	27.39
30	0.999	65.22	386.59	5,002	7,262	16,076	58,124	1258.48	285	26.90
50	0.333	76.40	384.54	6,983	0	24,383	60,140	6664.23	48	20.45
50	0.666	76.31	384.99	7,618	0	23,254	58,868	2496.08	150	20.14
50	0.999	76.50	385.36	8,271	0	21,955	58,037	1336.04	263	19.79
100	0.333	101.72	381.95	8,956	0	23,236	59,831	6847.82	42	18.36
100	0.666	101.65	382.57	9,773	0	21,862	58,640	2713.95	128	17.96
100	0.999	101.58	383.11	10,553	0	20,906	57,858	1511.50	208	17.57

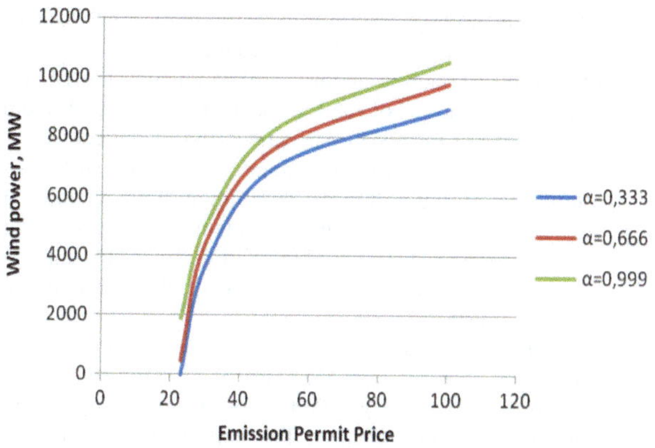

Fig. 6.2 Wind power with different emission permit prices and shares of customers on RTP (published with kind permission of © Elsevier B.J. 2012. All rights reserved)

The increase in wind power capacity produces decreased demand for mid-merit and peaker capacity. Because these capacities are based on fossil fuel this reduction is shown also as reduced emissions of carbon dioxide. The reduction is bigger the higher is the emissions price. The decrease in emission is strengthening as the share of the customers in RTP increases. This again confirms the argument of importance of RTP in reaching the EU set targets of mitigation in greenhouse gases.

Although some wind power is entering to the market as the carbon price increases, notably we do not reach the aggregate national target level for wind power (25,600 MW) even with very high carbon price. This is at least the case if wind power investors are risk averse and base their investment decisions on the lower bound of the resulting revenue (i.e., they assume that the wind is blowing during the lowest 2,500 demand hours when the market price is at lower level than on the hours located at peaker part of the load curve). Consequently, if we want to see that high level of installed wind power capacity also other support mechanisms are needed (in our paper Kopsakangas-Savolainen and Svento [10] we analyze different feed-in tariffs in reaching the goals for renewable energy sources).

References

1. Bunn DW, Fezzi C (2007) Interaction of European carbon trading and energy prices. Fondazione Eni Enrico Mattei working paper 123
2. Convery FJ (2009) Origins and development of the EU ETS. Environ Res Econ 43:391–412
3. Dijkstra BR, Manderson E, Lee T-Y (2011) Extending the sectoral coverage of an international emission trading scheme. Environ Res Econ 50:243–266
4. EC (2003) Directive 2003/87/EC of the European parliament and of the council of 13 Oct 2003 establishing a scheme for greenhouse gas emission allowance trading within the community and amending council directive 96/61/EC (text with EEA relevance). Official Journal L 275, 25/10/2003, pp 0032–0046
5. EC (2004) Directive 2004/101/EC of the European parliament and of the council of 27 Oct 2004 amending Directive 2003/87/EC establishing a scheme for greenhouse gas emission

allowance trading within the community, in respect of the Kyoto protocol's project mechanisms (text with EEA relevance)
6. EC (2008) EU action against climate change: the EU emissions trading scheme. Office for official publications of the European communities, Luxembourg. EC, 2008b. Directive 2008/101/EC of the European parliament and of the council of 19 Nov 2008 amending directive 2003/87/EC so as to include aviation activities in the scheme for greenhouse gas emission allowance trading within the community, Official Journal of the European Union L 8/3
7. EC (2009) Directive 2009/29/EC of the European parliament and of the council of 23 Apr 2009 amending directive 2003/87/EC so as to improve and extend the greenhouse gas emission allowance trading scheme of the community. Official Journal of the European Union L 140/63
8. Honkatukia J, Mälkönen V, Perrels A (2008) Impacts of the European emission trade system on finnish wholesale electricity prices. In: Gulli F (ed) Markets for carbon and power pricing in Europe. Edward Elgar Publishing, Cheltenham, pp 160–192
9. Kirat D, Ahamada A (2011) The impact of the European union emission trading scheme on the electricity-generation sector. Energy Econ 33:995–1003
10. Kopsakangas-Savolainen M, Svento R (2011) Promotion of market access for renewable energy. SSRN working papers. Available at SSRN: http://ssrn.com/abstract=1729723
11. Kossoy A, Ambrosi P (2010) State and trends of the carbon market 2010 report. Carbon Finance at the World Bank, Washington
12. NSW Government (2002) Amendment to the electricity supply act 1995 no. 94. New South Wales Government, New South Wales Legislation. http://www.legislation.nsw.gov.au/sessionalview/sessional/act/1995-94.pdf. Accessed 18 Jan 2012
13. NSW Government (2002) Amendment to the electricity supply (general) regulation 2001 (the regulation). New South Wales Government, New South Wales Legislation. http://www.treasury.nsw.gov.au/__data/assets/pdf_file/0004/5386/entire_doc.pdf. Accessed 18 Jan 2012
14. NZ Ministry for the Environment (2002) Climate change response act 2002 no. 40. New Zealand Legislation, NZ Parliamentary Counsel Office/Te TariTohutohu Paremata. http://www.legislation.govt.nz/act/public/2002/0040/37.0/DLM158584.html. Accessed 18 Jan 2012
15. NZ Ministry for the Environment (2008) Climate change response (emissions trading) amendment act 2008 no. 85. New Zealand Legislation, NZ Parliamentary Counsel Office/Te Tari Tohutohu Paremata. http://www.legislation.govt.nz/act/public/2008/0085/32.0/DLM1130932.html. Accessed 18 Jan 2012
16. Perdan S, Azapagic A (2011) Carbon trading: current schemes and future developments. Energy Policy 39:6040–6054
17. Point Carbon (2010) Carbon market trader EU. http://www.pointcarbon.com/trading. Accessed 18 Jan 2012
18. Reinaud J (2007) CO_2 allowances and electricity price interaction: impact on industry's electricity purchasing strategies in Europe. IEA information paper
19. RGGI (2005) Regional greenhouse gas initiative: memorandum of understanding, 20 Dec 2005. http://rggi.org/design/history/mou. Accessed 18 Jan 2012
20. RGGI (2005) Regional greenhouse gas initiative: amendment to memorandum of understanding, 31 Aug 2006. http://rggi.org/docs/mou_amendment_8_31_06.pdf. Accessed 18 Jan 2012
21. RGGI (2005) Regional greenhouse gas initiative: second amendment to memorandum of understanding, 20 Apr 2007. http://rggi.org/docs/mou_second_amend.pdf. Accessed 18 Jan 2012
22. RGGI (2010) Regional greenhouse gas initiative: an initiative of the Northeast and Mid-Atlantic states of the US. http://rggi.org/. Accessed 18 Jan 2012
23. Sijm JPM, Chen Y, Bakker SJA, Harmsen H, Lise W (2005) CO_2 price dynamics: the implications of EU emissions trading for the price of electricity. ECN report 2005, No. ECN-C-05-081
24. Sijm JPM, Chen Y, Donkelaar M, Hers JS, Scheepers MJJ (2006) CO_2 price dynamics: a follow-up analysis of the implications of EU emissions trading for the price of electricity. ECN report 2006, No. ECN-C-06-015

25. Tokyo Metropolitan Government (2008) Introduction of Tokyo's mandatory cap and trade system. Tokyo Metropolitan Government, Bureau of Environment, Environmental Policy Division, 25 June 2008. http://www.kankyo.metro.tokyo.jp/en/attachement/june25.pdf . Accessed 18 Jan 2012
26. Tokyo Metropolitan Government (2009) Tokyo cap-and-trade program: world's first Urban emission trading program, 26 May 2009. http://www.kankyo.metro.tokyo.jp/en/attachement/april1.pdf. Accessed 18 Jan 2012
27. Tokyo Metropolitan Government (2010) Tokyo cap-and-trade program: Japan's first mandatory emissions trading scheme. Emission cap and trade section, Urban and global environment division, Bureau of environment, Tokyo Metropolitan Government, March 2010. http://www.kankyo.metro.tokyo.jp/en/climate/cap_and_trade.html. Accessed 18 Jan 2012

Chapter 7
Efficiency of Electricity Distribution

7.1 Background

A basic feature of the restructuring of electricity markets has been the requirement to separate retail, generation, transmission, and distribution functions from vertically integrated companies to independent companies. This requirement is based on the argument that unbundling of the services makes it possible to gain greater efficiency through competition in retail and generation and reduces possibilities for cross-subsidization of generation activities with distribution activities. As we discussed in Chap. 2 this argument has a wide support and many countries have taken legal actions that necessitate unbundling of the vertically integrated firms. At the same time it is of course the case that vertical integration potentially creates benefits if generation and distribution necessitate substantial coordination across stages or if there are high transaction costs related to the use of intermediary markets.

This question of relative merits of unbundling of generation and distribution has not been extensively studied. Kaserman and Mayo [35], Kwoka [42], and Isaacs [31] have studied this question using a quadratic multi-stage cost function for U.S. electricity companies. Jara-Diaz et al. [33] studied Spanish, Fraquelli et al. [21] Italian, and Fetz and Filippini [20] Swiss electricity utilities. These studies do find economies of vertical integration to exist and they serve as a good reminder of this other side of the coin.

Following unbundling the energy power trade has been internationalized in the Nordic power region. However, based on natural monopoly features of distribution network the final delivery to users must be done by local distribution company. This is why we concentrate here on the Finnish distribution system. The efficiency questions we study generalize even though the concrete results of course are context-dependent.

The task of evaluating the efficiencies of the distribution utilities is basically a question of measuring relative efficiencies of different companies. Such a thing as

M. Kopsakangas-Savolainen and R. Svento, *Modern Energy Markets*, Green Energy and Technology, DOI: 10.1007/978-1-4471-2972-1_7,

an absolute efficiency is of course not possible to define or measure. Relative efficiencies can be measured with several methods. The two most common and widely used ones are Data envelopment analysis (DEA) and Stochastic frontier analysis (SFA). Both methods are based on the idea of constructing a frontier of the most efficient firms and then measuring inefficiencies of the firms outside of the frontier by the deviations from the frontier. DEA is a non-parametric method and SFA is a parametric method. This means that when using DEA any functional assumptions of the form of the frontier need not be made while this is necessary in the SFA.

DEA is the older of these methods and it originates from Farrell [15] and its current popularity lies much on Charnes et al. [10]. DEA has been widely used and as, e.g., Seiford [55] and Gattoufi et al. [23] show that it has decision-making relevance. At the same time it is, however, strongly criticized for its deterministic nature (e.g., [53]). This critique has been responded by developing the statistical base of DEA. It has been shown that DEA estimators in fact have a maximum likelihood interpretation [4]. The question of allowing genuine probabilistic randomness in the data is still very much an open question. The papers by Kuosmanen and Johnson [39] and Kuosmanen and Kortelainen [40, 41] start bridging this gap by demonstrating that DEA can be interpreted as a least-squares regression. They also show that the standard DEA model can be formulated as non-parametric least-squares regression under some constraints on the frontier shape and regression residuals. Kuosmanen and Johnson [39] develop a new method which they call *Corrected concave non-parametric least-squares* (C^2NLS) and they show that the estimates based on this method are consistent and asymptotically unbiased. Kuosmanen and Kortelainen [40, 41] take first steps in integrating truly stochastic inefficiency and noise terms into the non-parametric frontier analysis.

The regulator needs to have a reasonably correct information about the likely effects of their decision on the performance of the regulated utilities/institutions. During recent years research on measuring the performance of regulated firms has mainly focused on frontier efficiency. Frontier techniques in measuring efficiency have been seen superior because they use either statistical techniques or programing techniques in such a way that the effects of firm-specific differences and other exogenous factors are removed from affecting the performance ratios in order to obtain better estimates of the underlying performance of the firm (see e.g. [8]).

Regardless of the increasing amount of good quality research on efficiency measurement (see, e.g., [16–19, 22, 25, 26, 28, 32, 37, 41])[1] there is no clear consensus which is the best method for measuring frontier efficiency. However, the choice of the method may have significant impact on the policy conclusions which the regulatory authority is going to make based on the efficiency analysis.

[1] The semi-parametric frontier method introduced by Kuosmanen and Kortelainen [41] is perhaps the most promising method to solve most of the basic problems concerning SFA and DEA methods.

In this section we illustrate through a real life practical example the differences of the two main efficiency measurement methods used in electricity distribution regulation. For the regulator it is important that by using efficiency measurement methods it can diminish information asymmetries and have more reliable information on which to base its policies. In electricity distribution some countries have used methods based on DEA (see, e.g., [3, 22, 29, 30]) while others (an increasing number of countries) have used SFA (see, e.g., 17, 36]). Since the articles by Aigner and Chu [1], Aigner et al. [2], and Farrell [15] both the methodological development in frontier production function estimation and the resulting increase in the literature have been rapid.[2]

We use a real life regulation case by the Finnish energy market authority in regulating the Finnish electricity distribution network in order to illustrate the potential problems resulting from implementation of different efficiency measurement methods. We have chosen the Finnish case since the regulatory authority in Finland used simultaneously DEA and SFA for a 4-year regulatory period. The basic underlying regulatory model on which the authority used efficiency measurement was Rate of Return regulation. The basic regulatory procedure is such that in the beginning of each regulatory period the Energy Market Authority announces firm-specific regulatory decisions according to which the firm is allowed to set its network distribution price. The regulatory decision includes details on valuation of the invested capital, permitted rate of return, and goals for increasing efficiency. In the valuation of firm-specific efficiency the Finnish Energy Market Authority[3] used simultaneously both DEA[4] and SFA methods between the years 2008–2011. The reason for this parallel use is not clear. There have been many problems in parallel practical applications of these methods since they have given different efficiency results. In practice Energy Market Authority ended up in using the firm-specific efficiency scores calculated as the average of the DEA and SFA efficiency score. This operation mode was strongly criticised by the electricity companies. Consequently network operators made an appeal against this operational model to the Market Court. The main motivation for this section is to increase understanding of the differences and similarities of DEA and SFA methods in practical applications and give some advice whether the results of the methods are consistent with each other.

---

[2] For literature see, e.g., Barros and Peypoch [6], Battese and Coelli [7], Cornwell and Schmidt [13], Greene [24–26], Hjalmarsson, et al. [28], Kumbhakar and Lovell [38], Lovell [43], Pacudan and de Guzman [44], Pitt and Lee [45], Pombo and Taborda [47], Reichmann and Sommersguter-Reichmann [49], Rodríguez-Álvarez et al. [51], Schmidt and Sicles [53], and Seiford [54].

[3] The authority responsible for the electricity regulation in Finland is called Energiamarkkinavirasto (see www.energiamarkkinavirasto.fi for further information).

[4] The development procedure of the suggested DEA model used by Finnish Energy Market Authority is reported in Korhonen and Syrjänen [37]. It should be noted, however, that some of the suggestions (especially those concerning differences in operational conditions) have not been taken into account in practice.

In order to evaluate the consistency of these two methods we use the consistency condition criteria suggested by Bauer et al. [8].[5] The examined methods are DEA and two versions of the SFA. Our main results are that the consistency condition criteria are not fulfilled in many cases. According to the results both firm-specific efficiency scores and rank orders differ significantly from each other across various models. These differences are significantly bigger between the DEA and SFA models than among the two versions of the SFA model.

7.2 Consistency Conditions

Bauer et al. [8] proposed a set of consistency conditions that efficiency measures derived from different approaches should meet to be most useful for regulators. If different approaches give mutually inconsistent results the value of efficiency studies in practical regulatory applications is questionable. Although there is a relatively large literature on efficiency measurement there are only few studies which try to compare the efficiency measures resulting from different approaches. Studies of Burns and Weyman-Jones [9], Pollitt [46], Ray and Mukherjee [48], and Resende [50] are examples of the works which compare different approaches. However, neither of these studies makes consistency conditions as formal as Bauer et al. suggest (see e.g. Rossi and Ruzzier [52] on discussion of the studies comparing different approaches). According to Bauer et al. the efficiency estimates from the different approaches should be consistent in their efficiency levels, rankings, and identification of best and worst firms, consistent over time and with competitive conditions in the markets, and consistent with standard non-frontier measures of performance. These conditions can be presented as follows:

- Condition 1: The efficiency scores generated by different measuring approaches should have comparable distributional properties such as comparable means, standard deviations, etc.
- Condition 2: The different efficiency measuring techniques should rank the utilities in the approximately same order.
- Condition 3: The different efficiency measuring techniques should identify mostly the same utilities as the "best practice" and as the "worst practice".
- Condition 4: In order to be identified as consistent techniques all of the approaches should tend to identify the same utilities as relatively efficient or inefficient in different years. In other words different approaches should produce results that are reasonably stable over time.
- Condition 5: Competition condition in the markets and resulting efficiency scores generated by different approaches should be reasonably consistent.

[5] Resende [50] studied the robustness of different efficiency measures that can be used in the US telecommunications context when implementing incentive regulation. According to the results only moderate consistency across the different approaches was found.

- Condition 6: The measured efficiencies from different techniques should be consistent with the standard non-frontier performance measures (such as the cost/return ratio).

From these consistency conditions the first three ones measure mutual consistency of different approaches, whereas the last three ones evaluate how consistent different approaches are with reality or how believable they are. We analyze the first three mentioned in more detail because we are specifically interested in the mutual consistency of different approaches. If conditions 1–3 are satisfied the regulator can be confident that the efficiency scores resulting from analysis are correct and these scores can be used directly in benchmarking procedure (for example, in setting the X factor if the price cap regulation is going to be used). The countries' authorities responsible for regulation generally make the choice of the approach and it is more their task to evaluate the conditions from 4 to 6.

According to Rossi and Ruzzier [52] even if the condition (1) is not met, but conditions (2) and (3) are, the regulator has ordering of the firms by their efficiency levels and this information can be used, e.g., in setting the X factor in efficiency improvement target for the firms. Rossi and Ruzzier [52] and Bauer et al. [8] also argue that identifying the ordering of efficiency levels is usually more important on the regulatory decisions point of view than measuring the level of efficiency itself. Although neither the first nor the second condition is satisfied but the third consistency condition is met, it is possible for the regulator to use this information by publishing the results. This has been the approach used in the UK water and electricity sector (see [52]). The aim is to give firms incentive to improve their performance through public pressure.

There are a large variety of different tests which can be used in order to evaluate the fulfilment of different consistency conditions. In evaluating condition one, e.g., Kolmogorov–Smirnov test statistics can be used to study the distributional similarities of different methods. In studying the consistency of firm-specific efficiency scores some form of correlation test, such as Pearson test or Kendall test, can be used. According to condition 2 the different approaches should rank the utilities approximately in the same order. This can be studied by using, e.g., Spearman's rank correlations test. This test can be used in evaluating the fulfillment of condition 3.

Next we present the model specifications of the SFA and DEA evaluation methods used in 2008–2011 by the Finnish Energy Market Authority in electricity distribution regulation when evaluating the firm-specific efficiency.

7.2.1 Stochastic Frontier Analysis

Cost efficiency of each firm can be found out by relating its costs to the costs of firms closest to them but still included in the frontier, i.e., among the efficient ones.[6] A cost frontier can be expressed as

---

[6] Presentation of cost frontier models follows Kumbhakar and Lovell [38].

$$c_i \geq f(y_i, p_i; \beta), \quad i = 1, \ldots, N, \tag{7.1}$$

where $c_i = p_i^{'} x_i = \sum_k p_{ki} x_{ki}$, $k = 1,\ldots,m$ is the expenditure of firm i from production y_i using inputs x_{ki}, with prices p_{ki}. The vector of technology parameters β is to be estimated.

If data on input usage x_i of firm i is known it is possible to decompose cost efficiency into the cost of input oriented technical inefficiency and the cost of input allocative inefficiency. If data on input usage is not included in the data set this decomposition cannot be identified. The resulting inefficiency can thus be based either on non-optimal input combinations or on wrong choices from the expansion path.

The cost frontier in Eq. 7.1 is deterministic. It can be estimated using non-parametrical estimation methods such as DEA. Such a deterministic formulation ignores the fact that producers face random shocks that affect costs in ways not under control of producers. A stochastic cost frontier can be expressed as

$$c_i \geq f(y_i, p_i; \beta) \exp\{v_i\}, \tag{7.2}$$

where $[f(y_i, p_i; \beta) \exp\{v_i\}]$ is the stochastic cost frontier. This frontier now naturally consists of two parts: a deterministic part $f(y_i, p_i; \beta)$ common to all producers and a producer-specific random part $\exp\{v_i\}$ that captures the effects of random shocks on each producer.

To catch the firm-specific cost inefficiency another random term has to be included

$$c_i \geq f(y_i, p_i; \beta) \exp\{v_i\} \exp\{u_i\} \tag{7.3}$$

The composed error term $v_i + u_i$ now consists of two parts: v_i is a two-sided random-noise component and u_i is—when a cost frontier is in question—a non-negative cost inefficiency component. The composed error term is asymmetric being positively skewed since $u_i \geq 0$.

Assuming that the deterministic kernel takes the log-linear Cobb–Douglas form the stochastic cost frontier model can be written as

$$\begin{aligned} \ln c_i &\geq \beta_0 + \beta_y \ln y_i + \sum_k \beta_k \ln p_{ki} + v_i + u_i, \\ v_i &= N(0, \sigma_v^2), \quad u_i = N^+(0, \sigma_u^2). \end{aligned} \tag{7.4}$$

This cost frontier must be linearly homogenous in input prices and this can be attained through the reformulation

$$\begin{aligned} \ln\left(\frac{c_i}{p_{ji}}\right) &= \beta_0 + \beta_y \ln y_i + \sum_{k \neq i} \beta_k \ln\left(\frac{p_{ki}}{p_{ji}}\right) + v_i + u_i \\ v_i &= N(0, \sigma_v^2), \quad u_i = N^+(0, \sigma_u^2). \end{aligned} \tag{7.5}$$

As a summary we can say that a stochastic frontier cost function represents the minimum costs given the production technology, input variables, and output level. If the company fails to attain the cost frontier there are implications of the existence of technical/allocative inefficiency. The resulting estimated inefficiency scores represent the percentage deviation from a minimum level that would have been incurred if the company had operated as best practice (or cost-efficient) based on the data. This model can be estimated by the maximum likelihood and the cost efficiency of a firm i is given by Exp(u_i). In Chap. 8 we give examples on how to specify more elaborated versions of SFA. Here we keep to the basic form used by the Finnish regulator.

7.2.2 Data Envelopment Analysis

The DEA is still most widely used application in real life regulatory practices even though the use of SFA has increased during the last few years as discussed in this chapter. The basic model of DEA follows the presentation by Banker et al. [5] and Charnes et al. [10] and it can be expressed as:

$$\begin{aligned} &\min \theta \\ \text{s.t.}\quad &\theta x_{j0} - \sum_{i=1}^{n} \lambda_i x_{ji} \geq 0 \quad j = 1,\ldots,p, \\ &\sum_{i=1}^{n} \lambda_i y_{ki} \geq y_{k0} \quad j = 1,\ldots,r, \\ &\sum_{i=1}^{n} \lambda_i = 1, \\ &\lambda_i \geq 0 \quad i = 1,\ldots,n. \end{aligned} \tag{7.6}$$

This model is a input-oriented version of the DEA model and it thus minimizes the use of all the inputs given the level of output. The notation θ refers to the unit-specific efficiency score, y_{ki} refers to the output k produced by firm i, x_{ji} is the input j used by firm i, and λ_i refers to the variable weight of unit i in the reference point for the assessed unit. The convexity constraint, $\sum_{i=1}^{n} \lambda_i = 1$, enables the productivity to be dependent on the size of the unit in question. This minimization problem can be solved by using linear programing whereby a piece-wise linear "frontier" can be found. This frontier represents the minimum cost producing the given output. The solution thus gives a minimum feasible cost for each company and efficiency for each firm is estimated as its distance from this frontier.[7] This is the approach used in empirical DEA calculations below.

[7] See, e.g., Coelli et.al. [11] and Cooper et al. [12] for more specific presentation of the DEA.

7.3 Data

The data that we use in this book to study the efficiencies of the distribution utilities consists of 76 electricity distribution utilities in Finland. It covers the 6 year period from 1997 to 2002. The data, which is unbalanced panel data, is collected from the statistics of the Finnish Electricity Market Authority. Distribution utilities which are owned by industrial enterprises are excluded from this data. We use this quite old data because it covers the period after liberalization during which structural changes among the utilities did not emerge in a large scale. After this period there have been lots of mergers and other structural changes which make objective efficiency measurement problematic.

Table 7.1 gives the summary of descriptive statistics of the variables used in the analysis. We have used constant Euro prices in converting all money values to the year 1997 by using the retail price index.

Costs are expressed as average costs (c) calculated as total annual costs per kWh delivered and as total costs (C). Costs include the delivery to the final customers and the delivery to the networks. The costs of losses are excluded because of the lack of reliable data. Annual output (y) is measured in GWh and as can be seen from Table 7.1 it varies quite significantly since the range runs from very small local utilities to the relative large utilities operating on urban areas. The value of energy distributed (EV) is the amount of distributed electricity weighted by the average voltage-level-based distribution prices. CU is the total number of customers and as can be seen from Table 7.1 it ranges from 1,109 to 324,197 indicating the clear difference in the size of the distribution companies. Annual labor price p_l is calculated by dividing total annual labor cost by the average number of employees. The capital price p_k is calculated by dividing the annual capital expenditures by the value of capital stock. Total capital expenditure is calculated as residual costs. We have approximated the capital stock by the present value of the network. The present value of the network is calculated using the information of annual inventories and replacement value of the network. The price of the input power p_p is in most cases computational.[8] This is particularly the case when the distribution utility receives part of its delivered energy directly from the local generator and the other part outside of its own distribution network. Load factor (LF) is the ratio of the average load supplied during a designated period to the peak load occurring in that period, in kilowatts. Simply, the LF is the actual amount of kilowatt-hours delivered on a system in a designated period of time as

[8] The input price is computed when distribution utility receives part of its delivered energy outside of its own network and other part is received from local generator. The calculations are based on the market place payment, payment to the other companies' network, and on the relative share of the received power and delivered power. It is important to correct the input price since otherwise it distorts the cost structure of these companies.

Table 7.1 Descriptive statistics of Finnish distribution utilities 1997–2002

	Mean	S. D.	Minimum	Maximum
Total annual costs (C), (1,000 €)	7,096	12,316	221	108,758
Total annual costs(c) per kWh output (cents)	1.74	0.40	0.77	2.97
EV (1,000 €)	6,525	9,149	239	65,196
Annual output (y) in GWh	433.47	727.87	11.83	5,825.90
Number of customers (CU)	27,494	42,784	1,109	324,197
Load Factor (LF)	0.499	0.773E-01	0.191	0.866
Annual labour price (p_l) per employee (1,000 €)	28.39	7.75	8.14	53.00
Capital price (p_k) (1,000 €)	0.103	0.058	0.020	0.353
Price of input power (p_p) per kWh	0.36	0.14	0.09	1.06

opposed to the total possible kilowatt hours that could be delivered on a system in a designated period of time.[9]

As a parametric model we estimate cross-section and panel data versions of the stochastic frontier model. In the cross-section model we have used the average value of the six-year period of each variable. In the following we assume that the deterministic part of the SFA cost frontier takes the log-linear Cobb–Douglas form and linear homogeneity of cost frontier is attained by dividing costs by the price of the input power (p_p). In each model total annual costs (C) are explained by the energy value (EV), size of the network (AS), and number of customers (CU). We assume the random terms v and u to be normally and half normally distributed. The inefficiency term (u) is assumed to be time invariant.

The estimated model specification for cross-section and panel data versions of the SFA model is

$$\begin{aligned} \ln C_{it} &= \alpha + \beta_{EV} \ln EV_{it} + \beta_{AS} \ln AS_{it} + \beta_{CU} \ln CU + v_{it} + u_i, \\ v_{it} &= N(0, \sigma_v^2), \quad u_i = N^+(0, \sigma_u^2). \end{aligned} \tag{7.7}$$

The specification given in (7.1) can be used as such in DEA since DEA does not require any specific functional form. The value of energy (EV), number of customers (CU), and size of the service area (AS) are defined as outputs and total annual costs (C) are considered as an input.

[9] Utilities are generally interested in increasing LFs on their systems. A high LF indicates high usage of the system's equipment and is a measure of efficiency. High LF customers are normally very desirable from a utility's point of view. Using a year as the designated period, the LF is calculated by dividing the kilowatt-hours delivered during the year by the peak load for the year times the total number or hours during the year.

Table 7.2 Cost frontier parameters

	SFA(CS)		SFA(PD)	
	Coeff.	Std.er	Coeff	Std.er
Constant	2.409	0.310	2.879	0.259
lnEV	0.749	0.028	0.689	0.023
LnCU	0.251	0.033	0.333	0.034
LnAS	0.028	0.018	0.019	0.016
Log likelihood	24.531		114.774	

Table 7.3 Statistics of efficiency scores

	Mean	S. D.	Min	Max
SFA(CS)	0.81	0.07	0.57	0.94
SFA(PD)	0.77	0.12	0.42	0.99
DEA	0.77	0.16	0.37	1

7.4 Estimation Results and Consistency Condition for the SFA and DEA Model

Two versions of the SFA model and one version of the DEA model are estimated. The SFA(CS) refers to the SFA model applied to the cross-sectional data and SFA(PD) to the SFA model where panel data has been applied. Summary of the estimated efficiency scores are presented in Table 7.3 and cost frontier parameters for SFA models are presented in Table 7.2.

Estimation results show that coefficients of the frontier are significant and have expected signs. The size of the coefficients is relatively similar in cross-sectional and panel data versions of the model. The number of customers has somewhat smaller role in cost building in cross-sectional model than in panel data model. The opposite argument is true for the role of the area size and EV. Notice that the left hand side variable is now total costs and consequently we cannot see economics of scale in these estimations and thus the parameter for the volume variable is positive.

In interpreting the efficiency results the highest value 1 implies a perfectly efficient company and the difference from the efficiency value 1 and the firm-specific value in question tells us how much the firm can potentially save if it produces the output efficiently. The results indicate that on the average firms can improve their efficiency notably. Another clear observation is that according to all models there is large variation on the efficiency scores obtained by the estimation. One possible explanation is that electricity distribution companies operate in very heterogeneous environments in Finland and these model specifications do not take these environmental factors enough into account. The mean of efficiency scores are quite similar in all models. Notable is that the mean resulting from the SFA(CS) is somewhat higher than the ones obtained by SFA(PD) and DEA. The difference between the most efficient and most inefficient firm is bigger in DEA than resulting from the application of the SFA.

Table 7.4 Kolmogorov–Smirnov test statistics

	D-value[a]	P-value
SFA(CS)—SFA(PD)	0.053	0.982
SFA(CS)—DEA	0.072	0.807
SFA(PD)—DEA	0.079	0.713

[a] Maximum difference between the cumulative distributions

Table 7.5 Pearson correlation between efficiency scores, all firms

	SFA(CS)	SFA(PD)	DEA
SFA(CS)	1	0.977	0.515
SFA(PD)	0.977	1	0.499

Table 7.6 Pearson correlation between efficiency scores, 20 highest and 20 lowest firms

	SFA(CS) highest	Lowest	DEA highest	Lowest
SFA(CS)	1	1	0.773	0.973
SFA(PD)	0.978	0.986	0.820	0.962

Three of the consistency conditions recommended by the Bauer et al. [8] are analyzed next. First of these conditions requires that the efficiency scores generated by the different approaches should have comparable means, standard deviations, and other distributional properties. In order to test these observations statistically we use the Kolmogorov–Smirnov test for testing the equality of the efficiency distributions. The results of the Kolmogorov–Smirnov test are presented in Table 7.4. The null hypothesis for the test is that efficiency distributions among two specifications are the same.

According to the Kolmogorov–Smirnov test statistics the efficiency distributions are not statistically different from each other. However, the difference of the distribution is somewhat bigger when we compare the SFA and DEA models than in the case when panel data and cross-section versions of the SFA models are compared. From Table 7.3 it can be observed that both means and standard deviations are reasonably close to each other in all models. We also tested the correlation between firm-specific efficiency scores by using Pearson correlation test. The results can be seen from Table 7.5. The correlation among SFA models is very high and clearly higher than the correlation among the DEA and SFA models. We studied in more detail the correlation among the 20 most efficient firms (referred as "highest" in Table 7.6) and the 20 most inefficient firms (referred as "lowest" in Table 7.6). It seems that the correlation between two versions of the SFA models is very high also when we examine the 20 most efficient or 20 most inefficient firm. The result is quite opposite in the case where the correlation among SFA and DEA model is analyzed. However, as can be seen from Tables 7.5 and 7.6 correlation among DEA and SFA models is higher when only the 20 most efficient firms or 20 most inefficient firms are analyzed than when the hole set of efficiency results are analyzed.

Table 7.7 Spearman's rank correlation test, all firms

	SFA(CS)	SFA(PD)	DEA
SFA(CS)	1	0.982	0.488
SFA(PD)	0.982	1	0.473

Table 7.8 Spearman's rank correlation test, 20 highest and 20 lowest firms

	SFA(CS) highest	Lowest	DEA highest	Lowest
SFA(CS)	1	1	0.148	0.457
SFA(PD)	0.871	0.877	0.293	0.451

The second condition requires that the different approaches should rank the institutions in an approximately same order. We have tested this condition by ranking the firms from 1 to 76 according to their firm-specific efficiency scores (companies ordered according to SFA(CS) model) and then applying the Spearman's rank correlation test. The result for this test can be seen from Table 7.7. It seems that the two SFA models rank the firms almost identically. This is not true when we compare the rank order based on the SFA models and DEA models. DEA and SFA models rank the firms clearly in different order. The third consistency condition suggested by Bauer et al. requires that the different approaches should identify mostly the same institutions as "best practice" and as "worst practice". We have tested this condition by applying the Spearman's rank correlation to "best" 20 firms (rank according to SFA-CS model) and to the "worst" 20 firms. As can be seen from Table 7.8 the SFA and DEA models rank the "best" firms even more differently that was the case when the hole set of rank orders were analyzed.

These consistency results point clearly out the problems of benchmarking analysis which is that different models produce clearly different results. The results are very sensitive on the approach used (parametric versus non-parametric) and also in some scale on the model specification used. Notable is, however, that the sensitivity problem is not so clear when the efficiency estimation is carried out on the industry level than when the results are used on the firm level.

As a summary we can say that according to Bauer et al. [8] if the approaches produce consistent efficiency distributions, then the expected effects of the regulatory policies on the performances of the firms would be rather similar across the approaches. If the condition 2 is fulfilled, i.e., all approaches rank the firms in about the same order, then the policy decisions on which firms are required to improve their efficiency are the same regardless of the underlying measuring method. The condition 3 can be seen as a weaker condition of the condition 2 and if all methods rank the most efficient and inefficient firms about the same order then it is more possible that the regulatory authority does not make big mistakes on its policy instructions based on these efficiency results. Based on our results it can be said that the consistency conditions suggested here are not even closely fulfilled in the Finnish electricity distribution industry and hence the regulator should be very careful when using this kind of efficiency information in regulatory practices.

Table 7.9 Average distribution prices (cents/kWh) of Finnish utilities, 1.11.2011

	K1	K2	L1	L2	M1	M2	T1	T2	T3	T4
Average of whole land	8.63	7.44	5.56	5.09	7.10	5.48	4.31	3.91	3.06	2.86
Cheapest	4.27	3.59	3.56	3.38	3.56	3.59	2.37	2.44	1.29	1.23
Lowest quarter	7.53	6.26	5.00	4.57	6.07	4.78	3.74	3.39	2.64	2.58
Highest quarter	9.27	7.60	5.84	5.39	7.28	5.77	4.72	4.13	3.35	3.21
Most expensive	13.22	9.95	7.13	6.32	11.50	7.24	6.24	5.54	4.59	4.51

K1 Appartment, use of electricity max 2,000 kWh/year
K2 One-family house, use of electricity max 5,000 kWh/year
L1 One-family house, use of electricity max 18,000 kWh/year
L2 One-family house, use of electricity max 20,000 kWh/year
M1 Farm house, use of electricity max 10,000 kWh/year
M2 Farm house, use of electricity max 35,000 kWh/year
T1 Small enterprise, use of electricity max 150,000 kWh/year
T2 Small enterprise, use of electricity max 600,000 kWh/year
T3 Medium size enterprise, use of electricity max 2,000,000 kWh/year
T4 Medium size enterprise, use of electricity max 10,000,000 kWh/year

Especially, there can be problems if these firm-specific efficiency scores are used as such in the regulatory process.

Our results are in the line with the earlier studies from which we refer to the study of Farsi and Filippini [17], where they compare SFA, COLS, and DEA methods in analyzing Electricity distribution in Switzerland and they find that both efficiency scores and rank orders are different between the different approaches. This difference is clearer when the results from parametric and non-parametric techniques are compared than among the parametric models. Jamas and Pollitt [32] came to the same kind of conclusion when they analyzed 63 electricity companies from six different European countries by using SFA, COLS, and DEA methods. Estache et al. [14] studied the performance of the electricity distribution utilities in South America by using SFA and DEA and they also found only little support for the consistency of different conditions.

7.5 Stochastic Frontier Functional Form

In parametric estimation we have to make assumptions about the frontier functional form. Next we study this specification problem and we ask whether flexible functional forms outperform traditional functional forms. Specifically we test Cobb–Douglas and Translog specifications for the frontier.

Before we turn to estimation results we open one more basic question related to relative efficiency measurement. This relates to the heterogeneity of the operation environments of the distribution utilities. Some utilities operate in rural and some in urban environments thus facing different constraints when aiming at efficient operation and management. In Table 7.9 we present aggregate data of the used distribution prices of Finnish utilities (the prices exclude taxes). As can be seen from the table, there are big differences in these prices.

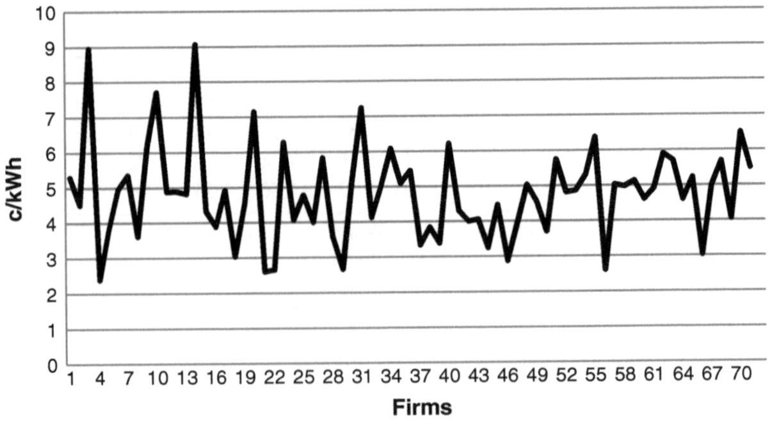

Fig. 7.1 The average costs (cents/kWh) normalized by input price (cents/kWh) of Finnish distribution utilities in 2002 (For data and description of input price see Table 7.1.)

These differences are understandable because the operation environments of the utilities are very different. Finland is sparsely inhabited (15 persons/km^2), with most of the population located in the south. In the sparsely inhabited areas the capacity requirement for the peak load is high compared to the average load of the network. In these areas capacity has to be sized according to high demand peaks which usually occur at the winter season when temperature falls occasionally very low (even at −50°C). In these sparsely inhabited areas also the number of customers is clearly lower than in the urban areas (which causes extra costs to the distribution companies which are beyond the managerial effort). These sparsely inhabited areas are also highly forested[10] with heavy winter snow-fall[11] which clearly affects the operational environment of the distribution utilities. These differences show up, e.g., in the load factors of the companies. The variation of the load factor is quite significant among different companies.

This vast heterogeneity of the utilities can also be seen from Fig. 7.1, where we present the normalized average costs of the Finnish utilities in 2002.

Obviously heterogeneity has to be taken into consideration in the estimations. Here we concentrate on the question of functional form of the frontier and in Chap. 8 we look more deeply the question of heterogeneity. Because heterogeneity is so obvious in electricity distribution industry we have however, tested functional forms also including basic heterogeneity characteristics. We have estimated five different SFA models with two functional form specifications which are Cobb–Douglas and Translog.

[10] The share of forested area in Finland is 68%. This means that Finland is the most forested nation in Europe.

[11] Heavy snow-fall causes frequent interruptions in electricity distribution.

The models we estimate are the true fixed effects (TFE) model proposed by Greene [25], the random effects (RE) model proposed by Pitt and Lee [45], and heterogeneity augmented models (REH) without observable heterogeneity explaining covariates in the frontier. We thus aim at reaching conclusions related to the frontier form using the basic models but adding basic heterogeneity explaining components to the model.

Assuming that the deterministic cost frontier takes the log-linear Cobb–Douglas form the stochastic cost frontier model can for the TFE be written as

$$\begin{aligned} \ln c_{it} &= \alpha_i + \beta_y \ln y_{it} + \beta_l \ln p_{lt} + \beta_k \ln p_{kt} + \tau T + v_{it} + u_{it} \\ v_{it} &= N(0, \sigma_v^2), \quad u_{it} = N^+(0, \sigma_u^2) \end{aligned} \tag{7.8}$$

and for RE model as

$$\begin{aligned} \ln c_{it} &= \alpha + \beta_y \ln y_{it} + \beta_l \ln p_{lt} + \beta_k \ln p_{kt} + \tau T + v_{it} + u_i \\ v_{it} &= N(0, \sigma_v^2), \quad u_i = N^+(0, \sigma_u^2) \end{aligned} \tag{7.9}$$

where c represents average costs, y is the quantity of delivered electricity measured in Gwh, p_l is labor price, p_k is price of capital, and T refers to the linear time trend. Linear homogeneity of cost frontier in input prices can be attained through the reformulation[12]

$$\begin{aligned} \ln\left(\frac{c_{it}}{p_{pit}}\right) &= \alpha + \beta_y \ln y_{it} + \beta_l \left(\frac{p_{lt}}{p_{pit}}\right) + \beta_k \left(\frac{p_{kt}}{p_{pit}}\right) + \tau T + v_{it} + u_i \\ v_{it} &= N(0, \sigma_v^2), \quad u_i = N^+(0, \sigma_u^2) \end{aligned} \tag{7.10}$$

We call the heterogeneity including models as REH(1), REH(2), and REH(3). We start by using the LF as the observable heterogeneity variable. We have selected this variable because it accounts well the differences in the operational environments among distribution utilities. The variation of the LF is quite significant among different companies. This variation is due to the fact that in the sparsely inhabited areas (located mostly in the northern and eastern parts of Finland) requirement for the peak load network capacity is high compared to the capacity required for the average load. It is also a fact that these areas, where the LF is relatively small, are the same areas where the land is highly forested (and thus also the snow burden is much higher in the forested areas than in the not forested areas).

LF is used for explaining observed heterogeneity in the inefficiency distribution as follows

REH(1) $u_i = N^+(\mu_i, \sigma_u^2)$, $\mu_i = \delta_0 + \delta_1 LF_i$,
REH(2) $\sigma_{ui}^2 = \exp(\delta_1 LF_i)$,
REH(3) $\sigma_{vi}^2 = \exp(\delta_1 LF_i)$.

---

[12] We use the price of the input power p_p as the normaliser.

The other specification of the cost function which we use is the Translog form of the cost frontier. We have excluded labor price and its quadratic form and quadratic form of the distributed kilowatts from this specification based on the estimation diagnostics. Cross-terms of these variables appear to be significant and thus they are included in the model. The Translog specification we use for the TFE and RE models is the following

$$\ln c = \alpha + \beta_y \ln y + \beta_k \ln p_k + \beta_{kk} \frac{1}{2} (\ln p_k)^2 + \beta_{yl} \ln y \ln p_l + \beta_{yk} \ln y \ln p_k + \beta_{lk} \ln p_l \ln p_k + \tau T + v + u, \tag{7.11}$$

where the firm index i and time subscript t have been left out.[13] For the models REH(1), REH(2), and REH(3) the heterogeneity specification is again based on the inefficiency distribution specifications presented above.

7.6 Frontier Functional Form Estimation Results

In Table 7.10 results for the estimations based on Cobb–Douglas specification are presented.[14] The dependent variable is average annual costs per kWh in 1997 cents. The numeraire input price is the input power price p_p.

The first observation on the estimation results is that all coefficients of the frontier are highly significant and have expected signs. Both price effects have positive signs in the all model specifications and the capital price effect is larger in absolute terms in all other than the TFE model. The high capital price estimates are understandable due to capital intensity of distribution networks. The sign of output (y) estimator is negative in all specifications which is expected since the explained variable is total average costs (cents per kWh). As the distributed quantity increases the unit costs decrease up to the point of minimum efficient scale. This is the first signal that there may be possibilities for companies to reduce their average costs by increasing output and that the firms may not operate at the point of minimum efficient scale. In order to study this question more closely we have also calculated the economies of scale. In the cost function framework there exists economies of scale if relative change in costs due to an increase in output is smaller than one. If the change is bigger than one there exists diseconomies of scale. In Tables 7.10 and 7.11 we show these values of economies of scale. Our results show that the values of economies of scale are less than one in all model specifications which suggests that distributors could lower their average costs

[13] In order to preserve comparability with the Cobb–Douglas models we use the same normalizations also here.

[14] We have used LIMDEP 9.0 [27] in all estimations.

Table 7.10 Cost frontier parameters: Cobb–Douglas specification

	TFE		RE		REH(1)		REH(2)		REH(3)	
	Coeff.	Standr. error	Coeff.	Standr. error	Coeff.	Standr. error	Coeff.	Standr. error	Coeff.	Standr. error
lny	−0.053	0.004	−0.107	0.008	−0.109	0.017	−0.103	0.008	−0.108	0.008
$\ln p_l$	0.427	0.016	0.292	0.009	0.276	0.010	0.292	0.009	0.278	0.010
$\ln p_k$	0.229	0.010	0.348	0.011	0.355	0.011	0.349	0.011	0.359	0.011
Constant			1.107	0.059	1.062	0.142	1.097	0.060	1.183	0.061
T	−0.026	0.005	−0.023	0.002	−0.023	0.002	−0.023	0.002	−0.025	0.002
Constant2					0.845	1.072				
lnLF					−2.207	0.530	−2.795	1.923	−1.764	0.319
$\sigma(v)$	0.188		0.082		0.081		0.083		0.043	
$\sigma(u)$	0.225		0.389		0.192		0.135		0.400	
λ[a]	1.199	0.212	4.718	1.338	2.363	0.421	1.629	1.533	9.339	5.130
N	419		419		419		419		419	
$\sigma^2(v)/\sigma^2(u)$	0.698		0.044		0.178		0.378		0.012	
BIC-criteria[b]	−382		−604		−616		−604		−608	
Economies of scale	0.947		0.893		0.891		0.897		0.892	

[a] $\lambda = \sigma(u)/\sigma(v)$

[b] BIC = −2*logL + Q*logN, where Q is the number of parameters

by increasing the output (assuming constant network size). The sign of the time estimate is negative. This indicates that there has been technological development which has decreased the total unit costs.

As can be seen from Table 7.10 TFE estimates are somewhat different from other models. One possible reason for this is the fact that in the TFE model all time invariant elements from the inefficiency are pushed to the constant terms of the model. It may be the case that this leads to underestimating the inefficiencies because part of the time invariant effects can be seen as true inefficiency. The range of the individual fixed effects parameters goes from 0.24 to 1.04 the average being 0.58. Compared to other terms of the model these are rather big which indicates that the fixed effects capture time invariant heterogeneity rather well. However, the variance of the frontier increases compared to the RE model. The Hausman test for the correlation between the individual effects and the explanatory variables is 130.3 indicating very strong correlation. This can be due to our way of explaining average costs with total kWh delivered. This high correlation indicates strong multicollinearity in the frontier part of the model which explains the increase of the frontier variance. This is also one explanation for the fact that the BIC-criteria do not favor the TFE model.

Considering the RE models the parameters between the RE and extended RE models are quite similar. This may indicate that even though the observed heterogeneity is at least partly explained by our variable lnLF the unobserved heterogeneity still appears as inefficiency. Because of this it is possible that the RE models overestimate the inefficiencies. In the estimations presented in Tables 7.10 and 7.11 firm-based observable heterogeneity in the model REH(1) is modeled by assuming that the mean of the inefficiency distribution is not zero but instead it is a function of heterogeneity explaining covariate lnLF: $E(u_i) = \delta_0 + \delta_1 \ln LF_i$. It can be shown that in order to have an inefficiency reducing result the sum $\delta_0 + \delta_1 \ln LF$ must be positive. As can be seen from column REH(1) in Table 7.10, our result confirms this.

The variance parameter of the underlying distribution of u_i, σ_u, is estimated as 0.389 in basic RE. In the expanded models REH(1) and REH(2) the counterparts are 0.192 and 0.135. These variances point out that some of the variation in the inefficiency in the original RE model can be explained as heterogeneity. The estimate of the frontier variance, σ_v, is almost unchanged for both REH(1) and REH(2) models. Third of the expanded models REH(3), where the heterogeneity component is included into the frontier variance, estimates the variance parameters clearly different from the other two expanded models.

It is also the case that those firms which have higher LF than the average are relatively more efficient. These are such firms which operate in less forested areas and which are average size firms. Firms whose LF is lower than the average seem to be less efficient and hence these firms may operate more cost efficiently if they raise their average load compared to the peak load. This can be obtained through increasing the average amount of the distributed kilowatts.

In Table 7.11 we present the cost frontier parameters for the Translog estimations. The first observation is that all coefficients of the frontier are again highly

Table 7.11 Cost frontier parameters: Translog specification

	TFE Model		RE Model		REH(1)		REH(2)		REH(3)	
	Coeff.	Standr. error	Coeff.	Standr. error	Coeff.	Standr. error	Coeff.	Standr. error	Coeff.	Standr. error
lny	−0.248	0.021	−0.273	0.017	−0.293	0.020	−0.269	0.016	−0.275	0.017
$\ln p_k$	1.628	0.093	1.271	0.085	1.167	0.085	1.263	0.085	1.247	0.090
$½*(\ln p_k)^2$	0.210	0.024	0.135	0.017	0.106	0.017	0.134	0.017	0.143	0.018
$\ln y*\ln p_l$	0.028	0.005	0.025	0.002	0.026	0.002	0.026	0.002	0.026	0.002
$\ln y*\ln p_k$	−0.048	0.005	−0.049	0.010	−0.049	0.010	−0.049	0.010	−0.049	0.010
$\ln p_l*\ln p_k$	−0.193	0.017	−0.109	0.012	−0.092	0.012	−0.107	0.012	−0.010	0.014
Constant			2.806	0.070	2.665	0.492	2.787	0.068	2.812	0.074
T	−0.027	0.005	−0.025	0.002	−0.025	0.002	−0.025	0.002	−0.027	0.002
Constant2					1.439	3.041				
lnLF					−2.128	0.540	−2.884	2.047	−2.370	0.332
σ (v)	0.170		0.085		0.083		0.085		0.035	
σ (u)	0.212		0.391		0.188		0.131		0.402	
λ	1.245	0.188	4.622	1.467	2.274	0.418	1.542	1.492	11.510	7.283
$\sigma^2(v)/\sigma^2(u)$	0.643		0.047		0.195		0.421		0.008	
N	419		419		419		419		419	
BIC-criteria	−422		−567		−586		−567		−578	
Econ. of scale	0.752		0.727		0.707		0.731		0.725	

significant and have expected signs. The estimator of capital price is positive and relatively big. This indicates the high capital intensity of the industry. Clear difference of this specification from the estimation results when we use Cobb–Douglas specification is that now the capital price effect is larger in TFE model than in all modifications of the RE models. The sign of output (y) estimator is again negative in all specifications as is expected. The output effect is very similar in all model specifications. This indicates that in all models firms operate in the decreasing part of the unit cost function. The sign of the time estimator is negative also in the Translog model. This again indicates that there has been technological development which has decreased total unit costs. The TFE estimators are somewhat different from other models also in this framework. This can again be explained by the feature that any unmeasured heterogeneity is placed to the cost function and this can produce a firm-specific shift of the cost function in the TFE model. In this specification the range for individual fixed effects is from 2.70 to 3.49 the average being 3.03. This again indicates that the individual fixed effects capture a lot of time invariant heterogeneity at the firm level. The Hausman test statistics is 223.5 indicating again strong correlation between effects and explanatory variables. Not surprisingly the frontier variance is bigger in TFE model than in RE model.

In the model REH(1) we again have a negative parameter for lnLF indicating $\delta_0 + \delta_1 \ln \mathrm{LF}$ to be positive. The variances of the underlying distribution of u_i, $\sigma_{u,}$ are somewhat similar to those estimated in the Cobb–Douglas specification. In basic RE model σ_u is estimated to 0.391. The counterparts of the models REH(1) and REH(2) are 0.188 and 0.131. This point out that some of the variation in the inefficiency in the original RE model can be explained as heterogeneity. The estimate of the frontier variance, σ_v, is almost unchanged for both REH(1) and REH(2) models compared to the basic RE model. Third of the expanded models REH(3) performs analogous to the one in the Cobb–Douglas specification.

Based on the BIC-criteria we can roughly say that among these models—independent of the cost frontier specification—the TFE models seem to fit the data worse than the RE models. The explanation is again the high correlation between the fixed effects and the explanatory variables. Among the RE models the model REH(1) fits the data best.

The values of economies of scale are smaller than one also in this specification indicating again that it would be possible for the companies to reduce their average costs by increasing the distributed quantity. It is also notable that according to this specification there exist more unexploited economies than was the case in Cobb–Douglas specification. It is, however, difficult for the regulator to give any exact practical rules for firms how they should increase their cost-effectiveness because companies' operational environments are very different. Again it seems clear that those firms which have higher LF than average can utilize their existing network best. Based on our data we can specify that these firms are average by size and they operate in relatively small or middle-sized urban areas.

Table 7.12 Statistics of inefficiency scores (Cobb–Douglas specification)

	TFE	RE	REH(1)	REH(2)	REH(3)
Minimum	0.105	0.952E-01	0.778E-01	0.956E-01	0.916E-01
Maximum	0.348	0.819	0.714	0.835	0.793
Mean	0.159	0.361	0.310	0.355	0.368
S.D. of $E[u_i \mid \varepsilon_i]$	0.297E-01	0.154	0.136	0.156	0.158

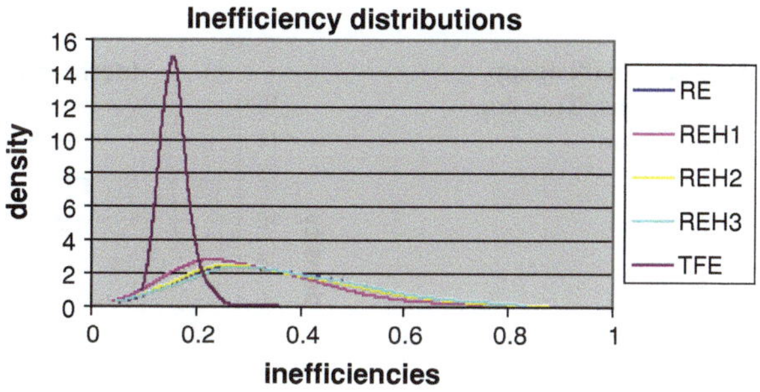

Fig. 7.2 Inefficiency distributions, Cobb–Douglas specification

7.7 Inefficiency Results

In Table 7.12 we present statistics of inefficiency scores for the Cobb–Douglas specification. The scores represent the percentage deviation from a minimum level that would have been incurred if the company had operated as best practice (or cost-efficient) based on our data. These scores have been calculated using the well-known Jondrow et al. [34] result which of course is here modified accordingly for each model specification.

In comparing different models it should again be stressed that TFE has distinctively different assumptions than all modifications of the RE models estimated. First difference is the assumption of time varying inefficiency over time. All RE models assume constant inefficiency over time. The second difference is that in TFE correlation between firm-specific fixed effects and explanatory variables is allowed. In the extended RE models the correlation is allowed through the lnLF variable. Third clear difference is that in all modifications of the RE models, i.e., RE, REH(1), REH(2), and REH(3), any unobserved firm-specific differences are interpreted as inefficiency. Given that in electricity distribution a considerable part of the unobserved heterogeneity is related to network characteristics and is very likely beyond the firm's own control, the inefficiency estimates can be overestimated in RE models. All these three distinguishing assumptions among TFE and RE models can be observed from our inefficiency estimates. The TFE model gives clearly lower inefficiency estimates than the other models. Since in the TFE model all of the invariant part of the inefficiency is pushed to the frontier this model may underestimate the inefficiencies. The mean inefficiency of the TFE model is 56% smaller than the mean inefficiency in the basic RE model. This high difference among mean inefficiency estimates may indicate that there are significant amounts of unobserved firm-specific factors which in the RE models appear in inefficiency scores.

Table 7.13 Statistics of inefficiency scores (Translog specification)

	TFE	RE	REH(1)	REH(2)	REH(3)
Minimum	0.879E-01	0.970E-01	0.643E-01	0.975E-01	0.931E-01
Maximum	0.321	0.789	0.661	0.803	0.749
Mean	0.150	0.365	0.283	0.358	0.372
S.D. of $E[u_i \vert \varepsilon_i]$	0.308E-01	0.149	0.124	0.151	0.153

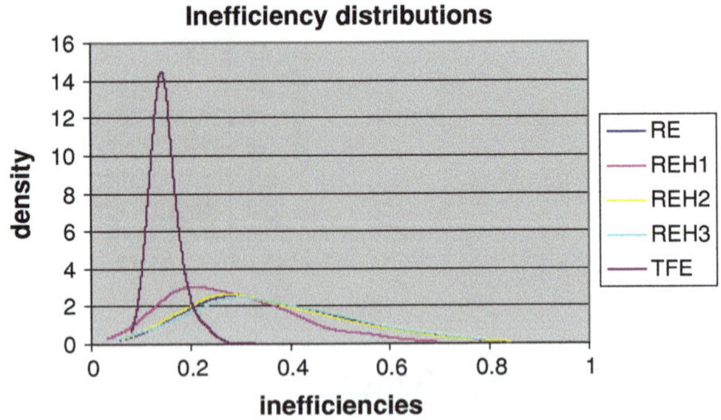

Fig. 7.3 Inefficiency distributions, Translog specification

When we compare the basic RE model to the extended RE models one observation to note is that mean inefficiency estimates diminish when the observed heterogeneity component is modeled into the mean or into the variance of the distribution of u_i. This result is clearest when we compare the basic RE model with the REH(1) model. According to the results some proportion of the variation in inefficiency seems to be explainable as heterogeneity in the mean. The difference of mean inefficiencies among RE and REH(2) models are very small which indicates that including heterogeneity component into the variance of u_i has insignificant impact to the inefficiency scores and is thus unable to explain the variation in inefficiency as heterogeneity but instead most of it still appears as inefficiency. In the REH(3) model the heterogeneity component is modeled into the variance of the frontier itself. This way of modeling heterogeneity operates to the opposite direction than the ones used in REH(1) and REH(2). If we assume that taking the heterogeneity into account should diminish inefficiency scores one can question the feasibility of this model.

In Fig. 7.2 the inefficiency distributions for the Cobb–Douglas models are presented. As can be seen also here the TFE model produces a clearly different inefficiency distribution compared to the RE models. All RE models seem to produce rather similar inefficiency distributions.

In Table 7.13 the inefficiency statistics for the Translog specification are presented. The results concerning inefficiency scores are quite similar to those of the Cobb–Douglas specification. Especially the REH(1) model seems to capture the observed heterogeneity more strongly than the corresponding Cobb–Douglas specification.

In Fig. 7.3 the distributions for the Translog specifications of $E[u_i|\varepsilon_i]$ are presented. These distributions show the same basic difference between TFE and

Table 7.14 Kolmogorov–Smirnov test statistics

	Cobb–Douglas		Translog	
	D-value	P-value	D-value	P-value
TFE–RE	0.3401	0.000	0.3842	0.000
TFE–REH(1)	0.3043	0.000	0.2876	0.000
TFE–REH(2)	0.3831	0.000	0.3294	0.000
TFE–REH(3)	0.3890	0.000	0.3842	0.000
RE–REH(1)	0.0656	0.052	0.1420	0.000
RE–REH(2)	0.0561	0.139	0.0680	0.038
RE–REH(3)	0.0800	0.009	0.0537	0.173
REH(1)–REH(2)	0.1074	0.000	0.0931	0.001
REH(1)–REH(3)	0.1289	0.000	0.1575	0.000
REH(2)–REH(3)	0.0585	0.110	0.0823	0.006

Table 7.15 Correlation between inefficiency ranks from different models (Cobb–Douglas specification)

	RE	REH1	REH2	REH3	TFE
RE	1				
REH1	0.999	1			
REH2	0.970	0.970	1		
REH3	0.991	0.990	0.961	1	
TFE	−0.055	−0.052	−0.014	−0.052	1

Table 7.16 Correlation between inefficiency ranks from different models (Translog specification)

	RE	REH1	REH2	REH3	TFE
RE	1				
REH1	0.990	1			
REH2	0.996	0.986	1		
REH3	0.985	0.979	0.980	1	
TFE	0.101	0.081	0.093	0.083	1

RE models as in the Cobb–Douglas case. The REH(1) model also produces an inefficiency distribution that seems to differ from other RE models.

In order to test these observations statistically we use the Kolmogorov–Smirnov test for testing the equality of the inefficiency distributions. The results of the Kolmogorov–Smirnov test are presented in Table 7.14. The null hypothesis for the test is that inefficiency distributions among two specifications are the same.

The inefficiency distributions are different from each other in all other cases than RE versus REH(1), RE versus REH(2), and REH(2) versus REH(3) in Cobb–Douglas frontier specifications and RE versus REH(2) and RE versus REH(3) in Translog specifications.

In practice the regulators use different benchmarking methods to rank companies according to their inefficiencies. The correlation matrixes based on Spearman's

Table 7.17 Correlation between inefficiency ranks from different model specifications (Cobb–Douglas versus Translog)

RE(C-D)–RE(Trans)	REH1(C-D)–REH1(Trans)	REH2 (C-D)–REH2(Trans)	REH3 (C-D)–REH3 (Trans)	TFE (C-D)–TFE(Trans)
0.983	0.921	0.982	0.983	0.838

correlation test between the ranks obtained by the inefficiency results from different models are presented in Tables 7.15 and 7.16. One observation to note is that the inefficiency ranks between all RE models show high correlation. Our analysis of inefficiency ranks indicates that especially the companies in the first and last 25% are ranked almost in the same order. The other clear observation is that when we use the TFE model in inefficiency estimation the resulting rank order differs considerably from the orders resulting by using any of the RE models. Actually the correlation between inefficiency ranks from TFE and any of the RE models are negative in Cobb–Douglas specification. Our inefficiency results also show that the 15 most efficient firms have high LF and are average by size.

It is clear that in both specifications for the frontier the TFE model ranks the firms very differently than the RE models and the RE models produce similar rankings. The rankings of TFE models are however, completely independent of the frontier specification, i.e., the Spearman correlation between the ranking of Cobb–Douglas TFE model and Translog TFE model is one (not shown here). This result again points out that fixed effects and RE models produce clearly different inefficiency results.

In Table 7.17 we show the rank correlations comparing rankings from different frontier model specifications. All correlations are very high showing that the frontier model specification is not as important for the rankings as it is the difference between fixed effects and RE models.

References

1. Aigner DJ, Chu SF (1968) On estimating the industry production function. Am Econ Rev 58:826–839
2. Aigner DJ, Lovell CAK, Schmidt P (1977) Formulation and estimation of stochastic frontier production function models. J Econometrics 6:21–37
3. Bagdadioglu N, Waddams PCM, Weyman-Jones TG (1996) Efficiency and ownership in electricity distribution: a non-parametric model of the Turkish experience. Energy Econ 18:1–23
4. Banker RD (1993) Maximum-likelihood, consistency and data envelopment analysis—a statistical foundation. Manage Sci 39:1265–1273
5. Banker RD, Charnes A, Cooper WW (1984) Some models for estimating technical and scale inefficiencies in data envelopment analysis. Manage Sci 30:1078–1092
6. Barros CP, Peypoch N (2009) An evaluation of European airlines' operational performance. Int J Prod Econ 122:525–533
7. Battese GE, Coelli T (1992) Frontier production functions, technical efficiency and panel data: with application to paddy framers in India. J Prod Anal 3:153–169

8. Bauer PW, Berger AN, Ferrier GD, Humphrey DB (1998) Consistency conditions for regulatory analysis of financial institutions: a comparison of frontier efficiency methods. J Econ Bus 50:85–114
9. Burns P, Weyman-Jones T (1996) Cost functions and cost efficiency in electricity distribution: a stochastic frontier approach. Bull Econ Res 48:41–46
10. Charnes A, Cooper WW, Rhodes E (1978) Measuring the efficiency of decision making units. Eur J Oper Res 2:429–444
11. Coelli T, Rao DSO, Battese G (1998) An introduction to efficiency and productivity analysis. Kluwer Academic Publishers, Boston
12. Cooper WW, Seiford LM, Tone K (1999) Data envelopment analysis: a comprehensive text with models, applications, references and DEA-solver software. Kluwer Academic Publishers, Boston
13. Cornwell C, Schmidt P (1996) Production frontiers and efficiency measurement. In: Matyas L, Sevestre P (eds.) The econometrics of panel data: a handbook of the theory with applications. Kluwer Academic Publishers, Dordrecht
14. Estache A, Rossi MA, Ruzzier CA (2004) The case for international coordination of electricity regulation: evidence form the measurement of efficiency in South America. J Regul Econ 25:271–295
15. Farrell MJ (1957) The measurement of productive efficiency. J R Stat Soc Ser A Stat Soc 120:253–281
16. Farsi M, Filippini M (2004) Regulation and measuring cost efficiency with panel data models application to electricity distribution utilities. Rev Ind Organiz 25:1–19
17. Farsi M, Filippini M (2005) A benchmarking analysis of electricity distribution utilities in Switzerland. CEPE working papers, p 43
18. Farsi M, Filippini M, Greene W (2005) Efficiency measurement in network industries: application to the Swiss railway companies. J Regul Econ 28:69–90
19. Farsi M, Filippini M, Greene W (2006) Application of panel data models in benchmarking analysis of the electricity distribution sector. Ann Public Co-Op Econ 77:271–290
20. Fetz A, Filippini M (2010) Economies of vertical integration in the Swiss electricity sector. Energy Econ 32:1325–1330
21. Fraquelli G, Piacenza M, Vannoni D (2005) Cost savings from generation and distribution with an application to Italian electric utilities. J Regul Econ 28:289–308
22. Försund FR, Kittelsen SAC (1998) Productivity development of Norwegian electricity distribution utilities. Resour Energy Econ 20:207–224
23. Gattoufi S, Oral M, Kumar A, Reisman A (2004) Content analysis of data envelopment analysis literature and its comparison with that of other OR/OM fields. J Oper Res Soc 55:911–935
24. Greene W (1993) The econometric approach to efficiency analysis. In: Fried HO, Lovell CAK, Schmidt SS (eds.) The measurement of productive efficiency: techniques and applications. Oxford University Press, Oxford
25. Greene W (2005) Reconsidering heterogeneity in panel data estimators of the stochastic frontier model. J Econometrics 126:269–303
26. Greene W (2005) Fixed and random effects in stochastic frontier models. J Prod Anal 23:7–32
27. Greene W (2007) LIMDEP version 9.0, econometric modelling guide, econometric software, Inc. Plainview, NY, USA
28. Hjalmarsson L, Kumbhakar SC, Heshmati A (1996) DEA, DFA and SFA: a comparison. J Prod Anal 7:303–327
29. Hjalmarsson L, Veiderpass A (1992) Efficiency and ownership in Swedish electricity retail distribution. J Prod Anal 3:7–23
30. Hjalmarsson L, Veiderpass A (1992) Productivity in Swedish electricity retail distribution. Scand J Econ 94:193–205
31. Isaacs J P (2006) Determining the regulatory impact on vertical economies in the electric utility industry. In: 4th annual international industrial organization conference, Boston

32. Jamasb T, Pollitt M (2003) International benchmarking and regulation: an application to European electricity distribution utilities. Energy Policy 31:1609–1622
33. Jara-Diaz SR, Martinez-Budria E, Ramos-Real FJ (2004) Economies of integration in the Spanish electricity industry using a multistage cost function. Energy Econ 26:995–1013
34. Jondrow J, Lovel CA, Materov I, Schmidt P (1982) On the estimation of technical inefficiency in the stochastic frontier production function model. J Econometrics 19:233–238
35. Kaserman DL, Mayo JW (1991) The measurement of vertical economies and the efficient structure of the electric utility industry. J Ind Econ 39:483–502
36. Kopsakangas-Savolainen M, Svento R (2008) Estimation of cost-effectiveness of the Finnish electricity distribution utilities. Energy Econ 30:212–229
37. Korhonen PJ, Syrjänen M (2003) Evaluation of cost efficiency in Finnish electricity distribution. Ann Oper Res 121:105–122
38. Kumbhakar S, Lovell K (2000) Stochastic frontier analysis. Cambridge University Press, Cambridge
39. Kuosmanen T, Johnson AL (2010) Data envelopment analysis as nonparametric least-squares regression. Oper Res 8:149–160
40. Kuosmanen T, Kortelainen M (2007) Stochastic nonparametric envelopment of data: cross-sectional frontier estimation subject to shape constraints. Economics discussion paper 46, University of Joensuu, Finland
41. Kuosmanen T, Kortelainen M (2011) Stochastic non-smooth envelopment on data: semi-parametric frontier estimation subject to shape constraints. J Prod Anal. doi:10.1007/s11123-010-0201-3
42. Kwoka JE (2002) Vertical economies in electric power: evidence on integration and its alternatives. Int J Ind Organ 20:653–671
43. Lovell CAK (1993) Production frontiers and productive efficiency. In: Fried HO, Lovell CAK, Schmidt SS (eds.) The measurement of productive efficiency: techniques and applications. Oxford University Press, Oxford
44. Pacudan R, de Guzman E (2002) Impact of energy efficiency policy to productive efficiency of electricity distribution industry in the Philippines. Energy Econ 24:41–54
45. Pitt M, Lee L (1981) The measurement and sources of technical inefficiency in Indonesian weaving industry. J Dev Econ 9:43–64
46. Pollitt M (1995) Ownership and performance in electric utilities: the international evidence on privatization and efficiency. Oxford University Press, Oxford
47. Pombo C, Taborda R (2006) Performance and efficiency in Colombia's power distribution system: effects of the 1994 reform. Energy Econ 28:339–369
48. Ray S, Mukherjee K (1995) Comparing parametric and non-parametric measures of efficiency: a re-examination of the Christensen green data. J Quant Econ 11:1
49. Reichmann G, Sommersguter-Reichmann M (2006) University library benchmarking: an international comparison using DEA. Int J Prod Econ 100:131–147
50. Resende M (2008) Efficiency measurement and regulation in US telecommunications: a robustness analysis. Int J Prod Econ 114:205–218
51. Rodrígues-Álvarez A, Tovar B, Trujillo L (2007) Firm and time varying technical and allocative efficiency: an application to port cargo handling firms. Int J Prod Econ 109:149–161
52. Rossi MA, Ruzzier CA (2000) On the regulatory application of efficiency measures. Util Policy 9:81–92
53. Schmidt P, Sickles RE (1984) Production frontiers and panel data. J Bus Econ Stat 2:367–374
54. Seiford LM (1994) A DEA bibliography (1978–1992). In: Charnes A, Cooper WW, Lewin AY, Seiford LM (eds.) Data envelopment analysis: theory, methodology, and application. Kluwer Academic Publisher, Boston
55. Seiford LM (1996) Data envelopment analysis: the evolution of the state of the art (1978–1995). J Prod Anal 7:99–137

Chapter 8
Observed Versus Unobserved Heterogeneity in Electricity Distribution

8.1 Background

The traditional models of stochastic frontiers have been extended so that firm-specific heterogeneity can better be taken into account. When the heterogeneity accounting literature started to develop it was first assumed that in the models time invariant parts represent inefficiency, whereas time variant parts can be seen as firm- or unit- specific heterogeneity. However, recently (see, e.g., [8–10]) this interpretation has radically changed. In recent papers it has been assumed that such parts of firm-specific effects which are not changing in time are mainly due to firm specific heterogeneity while the time variant part should be seen partly as inefficiency and partly as noise. Which one of these views is right is not an easy question. It is understandable that there are firm-specific heterogeneity factors which do not change in time and which are beyond the managerial effort. These should of course be interpreted as time-invariant heterogeneity. However, it is also possible that only part of the inefficiency is time variant. This is more likely to be the case if the industry under consideration is a (local) regulated monopoly and hence there may not exist full incentives to minimize costs.

If firm-specific heterogeneity is not accounted for it can create considerable bias in the inefficiency estimates. There have been, however, rapid developments in various forms of econometric methods during the past two decades which can, especially if we have panel data, identify observed and unobserved heterogeneity. The literature of panel data models in stochastic frontier analysis starts form Pitt and Lee [16] and is followed by Schmidt and Sickles [17] among others. During the past decade many authors (see, e.g., [11, 1]) have also included exogenous variables in the model to explain better the inefficiency component in the model.[1]

The much used basic model that assumes all heterogeneity to be explained by the covariates included in the frontier is the basic random effects (RE) specification

[1] See Kumbhakar and Lovell [15] for an extensive survey of stochastic frontier models.

M. Kopsakangas-Savolainen and R. Svento, *Modern Energy Markets*,
Green Energy and Technology, DOI: 10.1007/978-1-4471-2972-1_8,

proposed by Pitt and Lee [16]. In this model it is assumed that the firm-specific inefficiency (in proportional terms) is the same every year. There are some well-known problems connected to this model. One of them is that this model not only absorbs all unmeasured heterogeneity in the inefficiency term, but it also assumes that inefficiency is uncorrelated with included variables [9].

One way to overcome these problems is to model the observed heterogeneity[2] in the mean and/or variance of the distribution of inefficiency or to the variance of the distribution of the frontier error term (noise) as done in Chap. 7. We call this model specification again REH in the following.

One problem connected to the REH specification is that though the observed heterogeneity is now modeled out from the inefficiency distribution it does not recognize the unobserved heterogeneity which still remains in the inefficiency term. However, the second problem of the basic RE model is now reduced by allowing correlation between inefficiency explaining variables and frontier explaining variables. Another positive feature related to this model is that it enables a more precise estimation of the frontier.

Greene [9] proposes another extension to the RE model which is called the true random effects model (TRE). In fact this model has a predecessor in the literature. The model of Kumbhakar and Hjalmarsson [14] is essentially the same as the TRE model but the interpretation and estimation method differ substantially from that which Greene proposes. The TRE model is basically an ordinary RE model with two differences. Now the inefficiency measuring error term is time varying and a firm-specific time invariant random effect is added to represent the unobserved heterogeneity among firms. The inefficiency component now varies freely across time and firm. It is thus assumed that the unobserved differences across firms that remain constant over time are driven by unobserved characteristics rather than by inefficiency.

In the TRE model all time invariant inefficiency is interpreted as firm-specific heterogeneity and this part is now captured to the frontier and thus it does not appear as inefficiency anymore. This part of "inefficiency" is assumed to be caused by such time-invariant network characteristics (unobserved) which are beyond the firm's and manager's control and hence it is seen that this part is rather firm-specific heterogeneity than real inefficiency.

To summarize; in the TRE model any unobserved heterogeneity is taken into account but inefficiency distribution is not explained by observed heterogeneity variable, whereas in the REH model observed heterogeneity is taken into account but all unobserved heterogeneities still appears as inefficiency. It is, however, possible to take both the unobserved and observed heterogeneity into account at the same time by combining the models TRE and REH. We call this model the TREH model.

The basic fixed effects (FE) model can also be used in Stochastic Frontier Analysis. To overcome the well-known problems related to the basic FE model

---

[2] See Greene [8] for incorporating measured heterogeneity in the production function.

[17], especially the fact that in the FE model any time invariant unobserved heterogeneity appears in the inefficiency component, Greene [9] proposes a model where firm-specific constant terms are placed in the frontier and the inefficiency is time variant. Greene refers to this extended model as the "true fixed effects model" (TFE) to underline the difference with the FE framework commonly used. In the TFE model, FE represent the unobserved firm heterogeneity, not the inefficiency as in the original FE model. In other words it places unmeasured heterogeneity in the frontier and hence if the model is log linear it produces a neutral shift of the function specific to each firm. Greene shows, by using simulated samples, that although the FE may be largely biased, as far as the structural parameters and inefficiency estimates are concerned, the model performs reasonably well. The model can be fit by maximum likelihood.

Many empirical studies have already been done using these new models. Farsi and Filippini [2] studied cost-efficiency with panel data models in the Swiss electricity distribution utilities. In that paper, they utilized original RE and FE models and found that different model specifications could lead to different individual efficiency estimates. Kopsakangas–Savolainen and Svento [12, 13] utilized the variations of conventional RE models in measuring cost-effectiveness of Finnish electricity distribution utilities.

Farsi, Filippini, and Greene [4] applied stochastic frontier models in cost-efficiency measuring to the electricity distribution sector. In that paper they focus on three panel data models: GLS model Schmidt and Sickles [17], MLE model ([16], and the TRE model [9]). According to their results it is very important to model heterogeneity and inefficiency separately. In their paper (2005) Farsi, Filippini, and Greene studied network industries and compared different stochastic frontier models in a very comprehensive and detailed manner. It seems that the TRE model gives significantly lower inefficiency values than the other models they utilized. However, they point out a shortcoming of that model, namely that the firm specific heterogeneity terms are assumed to be uncorrelated with the explanatory variables. Farsi, Filippini, and Kuenzle [5] have found similar results connected to different model specifications in measuring regional bus companies' cost efficiencies. According to their results the TRE model seems to give much more plausible results than the other model specifications. In their paper concerning the efficiency of Swiss gas distribution sector Farsi, Filippini and Kuenzle [6] pointed out the importance of taking into account the output characteristics (such as customer density and network size) in the cost efficiency measuring process. Farsi, and Filippini [7] showed the advantages of recently developed panel data stochastic frontier models in the measurement of the cost-efficiency for multi-utility companies that operate in different sectors and that are characterized by a strong unobserved heterogeneity.

In this chapter we study the different ways how the firm-specific heterogeneity can be taken into account in the stochastic frontier models framework. Observed heterogeneity can be taken into account by incorporating firm-specific heterogeneity either in the estimated distribution of inefficiency or in the cost function itself. It is important to include observed firm-specific effects to the model because otherwise, e.g., Hausman test can reject the model because of the presence of such heterogeneity which is correlated with the regressors but not necessarily related to inefficiency in the model as such. Unobserved heterogeneity can be taken into account by randomizing some of the parameters of the model in which case it is assumed that this randomization captures all time-invariant unobserved heterogeneity.

We are especially interested in which kind of differences in inefficiency scores and firm rankings occur if we compare models that take into account only the observed heterogeneity to those that take into account also the unobserved heterogeneity. We take heterogeneity into account both through the inclusion of those effects in the cost function and in the mean or variance of the distribution of inefficiency (observed heterogeneity) and by randomizing some parameters of the stochastic frontier model (unobserved heterogeneity). We also estimate a combined model where we have randomized the frontier constant term and at the same time explained the mean of the inefficiency distribution by a covariate. We also estimate the true FE model, where the unobserved heterogeneity is represented by the individual FE.

8.2 Heterogeneity Augmented Versions of Stochastic Frontier Models

The different versions of the Stochastic Frontier models discussed above can be analytically expressed as follows. The basic fixed effect model can be presented as

$$\begin{aligned} c_{\mathrm{it}} &= \alpha_i + \boldsymbol{\beta}'\mathbf{x}_{\mathrm{it}} + v_{\mathrm{it}}, \quad v_{\mathrm{it}} = N(0, \sigma_v^2), \\ i &= 1, \ldots n\,; \quad t = 1, \ldots T. \end{aligned} \tag{8.1}$$

and the TFE model is of the form

$$\begin{aligned} c_{\mathrm{it}} &= \alpha_i + \boldsymbol{\beta}'\mathbf{x}_{\mathrm{it}} + v_{\mathrm{it}} + u_{\mathrm{it}}, \quad v_{\mathrm{it}} = N(0, \sigma_v^2), \quad u_{\mathrm{it}} = N^+\left(0, \sigma_u^2\right), \\ i &= 1, \ldots n\,; \quad t = 1, \ldots T. \end{aligned} \tag{8.2}$$

where c_{it} are the costs to be explained, x_{it} are the explaining variables, α and $\boldsymbol{\beta}$ are the parameters to be estimated, u_{it} is the half normally distributed inefficiency term with mean 0 and variance σ_u^2, and v_{it} is the normally distributed error term with variance σ_v^2. These models can be fit by maximum likelihood. The main feature of the TFE model is that it places the unmeasured heterogeneity in the cost function through firm-specific constants and thus with a log linear model, it produces a firm-specific shift of the cost function. It is assumed that the unobserved cost differences across firms that remain constant over time are not actually inefficiencies but rather they are some network-related unobserved characteristics.

The second model type that we use is based on the basic RE specification proposed by Pitt and Lee [16]. In this model it is assumed that the firm-specific inefficiency (in proportional terms) is the same every year as

$$\begin{aligned} c_{it} &= \alpha + \boldsymbol{\beta}'\mathbf{x}_{\mathrm{it}} + v_{\mathrm{it}} + u_i, \\ v_{\mathrm{it}} &= N(0, \sigma_v^2), \quad u_i = N^+(0, \sigma_u^2) \end{aligned} \tag{8.3}$$

where u_i and v_{it} are independent and, moreover, u_i is independent of x_{it}. Equation 8.3 can be estimated by maximum likelihood.

There are some recognized problems connected to this model. One of them is that this model not only absorbs all unmeasured heterogeneity in u_i, but it also assumes that inefficiency is uncorrelated with included variables [10]. This problem can be reduced through the inclusion of those effects in the mean and/or variance of the distribution of u_i or to the variance of the distribution of v_{it}. Another problem connected to the basic RE model is that the inefficiency term is time invariant.

While in the TFE model the unobserved heterogeneity is pushed into the cost function in the expanded random effect models the observed heterogeneity is resided either to the mean or to the variance of the inefficiency distribution or to the variance of the frontier. The first of the models which account observed heterogeneity in this way is the RE model extended by the inclusion of a heterogeneity component into the mean of the distribution of u_i. This model specification is called REH in the following. It can be written as

$$\begin{aligned} &c_{\mathrm{it}} = \alpha + \boldsymbol{\beta}'\mathbf{x}_{\mathrm{it}} + v_{\mathrm{it}} + u_{i,} \quad v_{\mathrm{it}} = N(0, \sigma_v^2), \quad u_i = N^+(\mu_i, \sigma_u^2), \\ &\mu_i = \delta_0 + \delta_1 h_i, \end{aligned} \tag{8.4}$$

where $\mathrm{h_i}$ is heterogeneity summarizing covariate explaining the mean of the inefficiency distribution and δ_0 and δ_1 are new parameters to be estimated. One problem connected to this specification is that though the observed heterogeneity is now modeled out from the inefficiency distribution it does not recognize the unobserved heterogeneity which still remains in u_i.

Another way of taking observed heterogeneity into account in the RE model is to include heterogeneity component into the inefficiency variance. We call this model specification as the REH (2) model. The basic interpretation of this model is very similar to that of REH, the only difference being that firm-specific inefficiency variances depend on the chosen cofactor. The REH (2) model is

$$\begin{aligned} &c_{\mathrm{it}} = \alpha + \boldsymbol{\beta}'\mathbf{x}_{\mathrm{it}} + v_{\mathrm{it}} + u_i, \quad v_{\mathrm{it}} = N(0, \sigma_v^2), \quad u_i = N^+(0, \sigma_{\mathrm{ui}}^2) \\ &\sigma_{\mathrm{ui}}^2 = \exp(\delta_1 h_i) \end{aligned} \tag{8.5}$$

where exp refers to the exponential distribution. It is also possible to include the observed heterogeneity component into the frontier variance. This model specification is called as the REH (3) model and it can be written as

$$\begin{aligned} &c_{\mathrm{it}} = \alpha + \boldsymbol{\beta}'\mathbf{x}_{\mathrm{it}} + v_{\mathrm{it}} + u_i, \quad v_{\mathrm{it}} = N(0, \sigma_{vi}^2), \quad u_i = N^+(0, \sigma_u^2) \\ &\sigma_{\mathrm{vi}}^2 = \exp(\delta_1 h_i). \end{aligned} \tag{8.6}$$

In this specification, the observed heterogeneity affects the expected inefficiencies through the variance of the frontier itself. It is important to note that all REH models are based on different ways of modeling the observed heterogeneity and it is difficult to evaluate beforehand any kind of superiority of these models in inefficiency measurements. One must also be careful in making interpretations, as the unobserved heterogeneity still remains in the inefficiency distributions.

The shortcoming of the RE models presented above is that they do not take unobserved heterogeneity into account. One way to include unobserved heterogeneity is to random parameterize these RE models. Greene proposed [9, 10] a random parameterized model which he calls the TRE model. This model is of the type

$$
\begin{aligned}
c_{\text{it}} &= (\alpha + \mathrm{w}_i) + \boldsymbol{\beta}'\mathbf{x}_{\text{it}} + v_{\text{it}} + u_{\text{it}}, \\
v_{\text{it}} &= N(0, \sigma_v^2), \quad u_{\text{it}} = N^+(0, \sigma_u^2)
\end{aligned}
\tag{8.7}
$$

where $\mathrm{w_i}$ is a normally distributed random variable. The estimation is based on simulation with draws from the normal distribution for w_i. Now w_i captures the unobserved heterogeneity. The observed heterogeneity augmented version of the true random effects model (TREH) is naturally one that includes heterogeneity explaining covariates in the frontier or in the efficiency distribution. This model specification can be written as

$$
\begin{aligned}
&c_{\text{it}} = c_{\text{it}} = (\alpha + \mathrm{w}_i) + \boldsymbol{\beta}'\mathbf{x}_{\text{it}} + v_{\text{it}} + u_{\text{it}}\, v_{\text{it}} = N(0, \sigma_v^2), \quad u_{\text{it}} = N^+(\mu_i, \sigma_u^2), \\
&\mu_i = \delta_0 + \delta_1 h_i,
\end{aligned}
\tag{8.8}
$$

The main features of the models described above are summarized in Table 8.1.

8.3 Empirical Versions of Heterogeneity Augmented SFA Models

Next, we shall turn into looking at the question of observed versus unobserved heterogeneity empirically. We shall do this by including observed heterogeneity explaining covariates in the frontier and allowing some of the parameters to be randomized. We estimate five modifications of the SFA models by using Cobb-Douglas specifications for the frontier. Assuming that the deterministic cost frontier takes the log-linear Cobb–Douglas form the empirical specification of the stochastic cost frontier RE model can, using our data,[3] be written as

$$
\begin{aligned}
&\ln c_{\text{it}} = \alpha + \beta_{\text{y}} \ln \mathrm{y}_{\text{it}} + \beta_{\text{LF}} \ln \text{LF}_{\text{it}} + \beta_{\text{CU}} \ln \text{CU}_{\text{it}} + \beta_l \ln p_{\text{lit}} + \beta_k \ln p_{\text{kit}} + \beta_T T + v_{\text{it}} + u_i \\
&v_{\text{it}} = N(0, \sigma_v^2), \quad u_i = N^+(0, \sigma_u^2), \quad i = 1, \ldots, 76,
\end{aligned}
\tag{8.9}
$$

[3] The data we use here is the same as that described in Chap. 7.

Table 8.1 Summary of the heterogeneity augmented SFA models

Model	Observed heterogeneity	Unobserved heterogeneity
Random effects (RE)	• Firm-specific observed factors in the frontier	• Not included in the model
Heterogeneity extended random effects (REH)	• Firm-specific observed factors in the frontier • Heterogeneity in the mean of inefficiency distribution or in the variances of either inefficiency or frontier	• Not included in the model
True random effects (TRE)	• Firm-specific observed factors in the frontier	• Time-invariant random component captures firm-specific unobserved heterogeneity
True fixed effects (TFE)	• Both with firm-specific observed factors in the frontier and without them	• Time-invariant fixed component captures firm-specific unobserved heterogeneity
Heterogeneity extended true random effects (TREH)	• Firm-specific observed factors in the frontier • Heterogeneity in the mean of inefficiency distribution	• Time-invariant random component captures firm-specific unobserved heterogeneity

where T again refers to time which is expected to capture possible technical change. We have used the variables Load Factor (LF) and number of customers (CU) as the indicators of observable heterogeneity in the frontier. We include the LF in the frontier based on the idea that it represents the utilization ratio of network capacity and thus it captures the technical features of the network. Although these variables are time variant the actual variation within one firm is very small. In our situation the explained variable is average cost so that these variables capture well the observed firm-specific heterogeneity.

The next estimated model is the RE model extended by the inclusion of a heterogeneity component into the mean of the distribution of u_i (REH).

$$\begin{aligned}
&\ln c_{\mathrm{it}} = \alpha + \beta_{\mathrm{y}} \ln y_{\mathrm{it}} + \beta_{\mathrm{LF}} \ln \mathrm{LF}_{\mathrm{it}} + \beta_{\mathrm{CU}} \ln \mathrm{CU}_{\mathrm{it}} + \beta_l \ln p_{\mathrm{lit}} + \beta_k \ln p_{\mathrm{kit}} + \beta_T T + v_{\mathrm{it}} + u_i \\
&v_{\mathrm{it}} = N(0,\sigma_v^2),\ \ u_i = N^+(\mu_i,\sigma_u^2), \\
&\mu_i = \delta_0 + \delta_1 \ln LF_i, \quad i = 1,\ldots 76.
\end{aligned} \tag{8.10}$$

The heterogeneity variable we again use is the LF. It is clearly higher in urban areas than in rural areas. In these sparsely inhabited areas the number of customers also is clearly lower than in the urban areas (which causes extra costs to the distribution companies which are beyond the managerial effort).

The next estimated model, TRE, is the random parameter version of the RE model. Now also the inefficiency term (u) is time variant. In the TRE model a firm-specific random constant term is used

$$\begin{aligned}\ln c_{\text{it}} &= (\alpha + \text{w}_i) + \beta_{\text{y}} \ln \text{y}_{\text{it}} + \beta_{LF} \ln \text{LF}_{\text{it}} + \beta_{CU} \ln \text{CU}_{\text{it}} \\ &\quad + \beta_l \ln p_{\text{lit}} + \beta_k \ln p_{\text{kit}} + \beta_T T + v_{\text{it}} + u_{\text{it}} \\ v_{\text{it}} &= N(0, \sigma_v^2), \quad u_{\text{it}} = N^+(0, \sigma_u^2), \qquad i = 1, \ldots, 76,\end{aligned} \tag{8.11}$$

We also estimate here the true fixed effect model (TFE)

$$\begin{aligned}&\ln c_{\text{it}} = \alpha_i + \beta_{\text{y}} \ln \text{y}_{\text{it}} + \beta_{\text{LF}} \ln \text{LF}_{\text{it}} + \beta_{\text{CU}} \ln \text{CU}_{\text{it}} + \beta_l \ln p_{\text{lit}} + \beta_k \ln p_{\text{kit}} + \beta_T T + v_{\text{it}} + u_{\text{it}} \\ &v_{\text{it}} = N(0, \sigma_v^2), \quad u_{\text{it}} = N^+(0, \sigma_u^2), \quad i = 1, \ldots, 76,\end{aligned} \tag{8.12}$$

The last estimated model is the TREH model which is the combined model of the models REH and TRE. The idea of this combined model is that it can take both the observed and unobserved heterogeneity into account at the same time

$$\begin{aligned}\ln c_{\text{it}} &= (\alpha + \text{w}_i) + \beta_{\text{y}} \ln \text{y}_{\text{it}} + \beta_{\text{LF}} \ln \text{LF}_{\text{it}} + \beta_{\text{CU}} \ln \text{CU}_{\text{it}} + \beta_l \ln p_{\text{lit}} \\ &\quad + \beta_k \ln p_{\text{kit}} + \beta_T T + v_{\text{it}} + u_{\text{it}} \\ v_{\text{it}} &= N(0, \sigma_v^2), \; u_{\text{it}} = N^+(\mu_{\text{it}}, \sigma_u^2), \\ \mu_{\text{it}} &= \delta_0 + \delta_1 \ln \text{LF}_{it}, \qquad i = 1\ldots, 76.\end{aligned} \tag{8.13}$$

In Table 8.2 results for these estimations are presented. The first observation on the estimation results is that all covariate coefficients of the frontier are highly significant[4] and have expected signs. Both price effects have positive signs in all model specifications and the capital price effect is larger in absolute terms in all other models than the TFE model. The high capital price estimates are understandable due to capital intensity of distribution networks. The sign of output (y) estimator is negative in all specifications which is expected since the explained variable is total costs per kWh. As the distributed quantity increases the unit costs decrease up to the point of minimum efficient scale. Also the sign of the time estimate is negative. This indicates that there has been technological development which has decreased the total unit costs.

The variance parameter of the underlying distribution of u_i, σ_u, is estimated as 0.353 (see Table 8.3) in basic random effects model (RE). In the extended version of the RE (REH) as well as in the randomized version TRE and TFE and in the combined model TREH counterparts are 0.150, 0.096, 0.101, and 0.106. These point out that some of the variations in the inefficiency in the original RE model can be explained as heterogeneity. Based on this notification we can expect the estimated inefficiencies to diminish. According to BIC-criteria it seems that the model which accounts both the observed and unobserved heterogeneity at the same time, i.e., the combined model TREH fits the data best.

In Table 8.3 we present statistics of inefficiency scores. The scores represent the expected percentage deviation from a minimum level that would have been

[4] Except LF in the RE model.

Table 8.2 Cost frontier parameters of models 1–5

	RE		REH		TRE		TFE		TREH	
	Coefficients	Standard error	Coefficients	Standard error	Coefficients	Standard error	Coefficients	Standard error	Coefficients	Std.er
Constant	−1.614	0.272	−2.326	0.126 + 07	−1.484	0.049			−1.079	0.069
lnY	−0.647	0.053	−0.657	0.050	−0.703	0.010	−0.595	0.018	−0.670	0.013
LnCU	0.584	0.054	0.603	0.052	0.644	0.010	0.569	0.016	0.582	0.014
LnLF[a]	−0.057	0.050	−0.020	0.048	−0.032	0.014	−0.352	0.033	0.034	0.019
LnLF			−2.550	0.675						
$\ln p_l$	0.297	0.008	0.288	0.007	0.300	0.004	0.445	0.014	0.321	0.006
$\ln p_k$	0.386	0.009	0.394	0.009	0.402	0.003	0.277	0.012	0.397	0.005
T	−0.014	0.002	−0.014	0.002	−0.015	0.001	−0.017	0.005	−0.016	0.001
Scale parameters for distributions[b]					0.193	0.003			0.151	0.003
Log likelihood	389.40		413.52		416.79		302.60		421.73	

[a] In the model REH this second LnLF refers to the third equation in model (2)
[b] Scale parameter for distributions of random parameters

Table 8.3 Statistics of inefficiency scores

	RE	REH	TRE	TFE	TREH
Minimum	0.972–01	0.419–01	0.117–01	0.575–01	0.102–01
Maximum	0.782	0.481	0.450	0.142	0.445
Mean	0.327	0.141	0.737–01	0.775–01	0.808–01
Standard deviation of $E[u_i\|\varepsilon_i]$	0.130	0.738–01	0.470–01	0.948–02	0.600–01
σ(v)	0.068	0.067	0.032	0.165	0.023
σ(u)	0.353	0.150	0.096	0.101	0.106

incurred if the company had operated as best-practice (or cost efficient) based on our data.

These basic statistics clearly show that all heterogeneity (either observed, unobserved or both) accounting models capture the firm-specific heterogeneity into the cost frontier allowing the inefficiency distribution move to the left and become more concise. Compared with the models without observed heterogeneity explaining covariates in the frontier (Table 7.12) we clearly see how the inefficiency scores for the comparable models TFE, RE, and REH diminish. Also noticable is that the distribution of the frontier in randomized specifications is more concise. Another clear observation again is that TFE produces here too clearly different inefficiency scores than either the basic RE model or the random parameterized versions of the RE model. The difference between the basic RE model inefficiency scores and those which the TFE model produces can be explained by the clearly different model assumptions. The first difference is the assumption of time-varying inefficiency. Both the RE and the REH models assume constant inefficiency over time. The second difference is that in the TFE correlation between firm-specific effects and explanatory variables is allowed. This is not the case for the basic RE model. The third clear difference is that in the basic RE model any unobserved firm-specific differences are interpreted as inefficiency. Given that in electricity distribution a considerable part of the unobserved heterogeneity is related to network characteristics and is very likely beyond the firm's own control, the inefficiency estimates can be overestimated in the RE models. All these three distinguishing assumptions between the TFE and RE models can be observed from our inefficiency estimates. It is notable that the variance of the frontier in the TFE model is rather big (0.165), which shows that the model does not produce robust estimates for the frontier. It is possible that the TFE is moving "too much" of the random variation into the noise compared to the other models. This can be due to the rather short panel of the data or insufficient number of observations.

When we compare the basic random effect model to the random parameterized versions of the RE model one observation to note is that mean inefficiency estimates clearly diminish. This can be explained by the fact that in the random parameterized models unobserved heterogeneity is not appearing as inefficiency. However, taking into account the fact that the firms in question are local monopolies it is possible that they do not operate as efficiently as possible and consequently part of the time-invariant inefficiency (now assumed to be due to

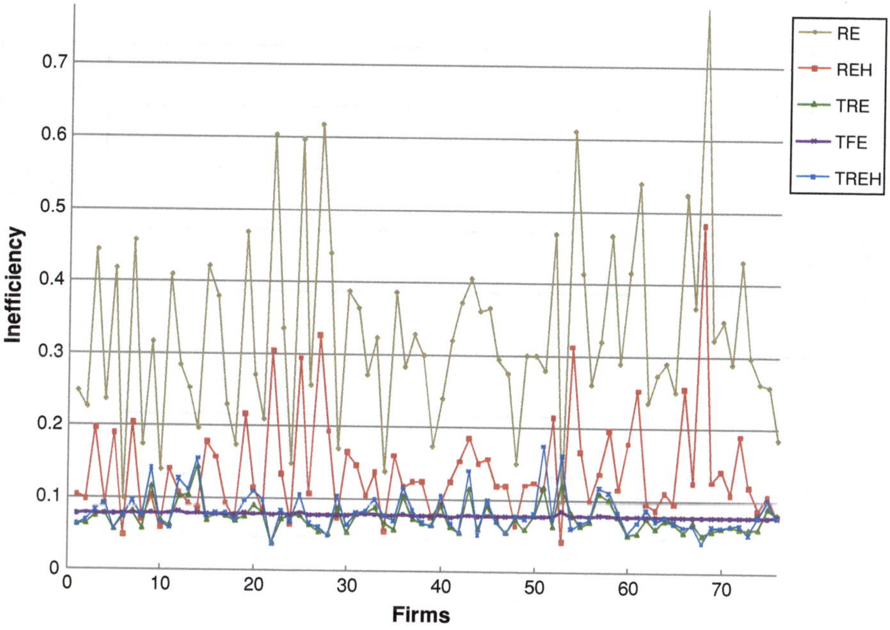

Fig. 8.1 Firm-specific expected mean inefficiencies

firm-specific unobserved heterogeneity) may be due to inefficient management and hence the model TRE may underestimate the inefficiency scores.

Also, the inefficiency scores among the TFE and random parameterized versions of the RE model differ somewhat. Although the maximum inefficiency score is clearly smaller in the TFE model the mean inefficiency is higher in the TFE model than in the TRE model. However, the mean of inefficiency is clearly closer to each other between the TFE model and the random parameterized version of the RE model than between the basic RE model and its heterogeneity accounting versions.

In Fig. 8.1 we plot the firm-specific expected mean inefficiencies. Figure 8.1 confirms the basic information about Table 8.3 namely that the models RE and REH cause the inefficiencies to be relatively bigger and more volatile than the other models. Also notable is that the behavior of firm-specific inefficiencies between the RE and the REH is quite similar which indicates positive firm-specific inefficiency score correlations. The models TRE, TFE, and TREH produce the expected firm-specific mean inefficiencies to be clearly smaller and less volatile. The behavior of firm-specific inefficiencies is quite similar among these models but for some firms clearly different from the behavior of inefficiencies based on the models RE and REH. This clearly points to the importance of modeling unobserved heterogeneity explicitly.

In Fig. 8.2 the inefficiency distributions of all models are presented. The deviation of the distributions of different models is clear. One notification to make is also that the kurtosis of inefficiency distribution of the TFE model is clearly higher than those in other model specifications.

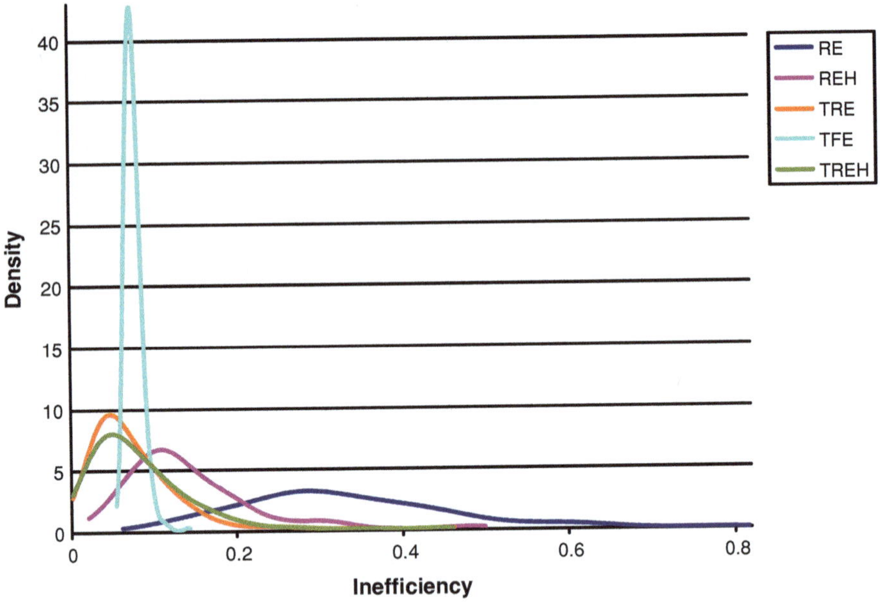

Fig. 8.2 Inefficiency distributions

Table 8.4 The Spearman correlations of the inefficiency rankings

	RE	REH	TRE	TFE	TREH
RE	1				
REH	0.962	1			
TRE	−0.332	−0.300	1		
TFE	−0.085	−0.057	0.217	1	
TREH	−0.280	−0.334	0.309	0.069	1

In practice, the regulators use different benchmarking methods to rank companies according to their inefficiencies. The correlation matrixes based on Spearman's correlation test between the ranks obtained by the inefficiency results from different models are presented in Table 8.4. One observation to note is that the inefficiency ranks between the basic RE model and all models other than REH are negatively correlated. The correlation between the RE model and its extended form REH are positive and close to one which indicates that these models rank the firms very similarly.

When we investigate the ranks firm by firm we notice that when the unobserved heterogeneity is taken into account the rank of the firms which are located in sparsely inhabited rural areas with long distribution distances increases and the ranks of those in relatively big cities decreases. Based on this observation it seems that these models produce rank orders which take into account such heterogeneity factors that are beyond the control of the firm or its managers and hence are such factors that should not be considered as inefficiency in regulatory benchmarking.

The problem associated with the Spearman test statistics is that it tests the monotonic relation between two variables. This relation exists when any increase in one variable is invariably associated with either an *in*crease or a *de*crease in the other variable. This means that this test does not recognize the distance among two variables. This potentially distorts these results.

Interestingly, the inefficiency scores for some firms seem to correlate negatively between these two groups of models, i.e., RE and REH versus TRE, TFE, and TREH. The main interest here was to study the different ways of how the firm-specific heterogeneity can be taken into account in the stochastic frontier model framework. We look at the potential advantages of heterogeneity extended stochastic frontier models over conventional RE models in cost-efficiency measurement. Especially, we are interested in how the inefficiency estimates change when we use random parameter models instead of conventional RE models. We have applied a basic RE model, one version of extended RE model where observed heterogeneity is captured by explaining the mean of the inefficiency with the LF covariate and two random parameterized stochastic frontier models (these models are assumed to take the unobserved heterogeneity into account) from which the first one is the so-called TRE model and the second is a model that combines the true random effect model with the model that explains the mean of the inefficiency distribution by some covariate. We also estimate the TFE model where firm-specific constants capture unobserved heterogeneity. Our data consists of 76 regional distribution utilities which vary significantly if measured by output as well as by the operative environment. Our basic result is that random parameter estimation of stochastic cost frontiers produce clearly smaller inefficiency estimates than the basic RE model or its extended version. The inefficiency estimates produced by the heterogeneity accounting version of the basic RE model are also clearly smaller than the one resulting from basic RE model. Note, however, that even though both ways of accounting heterogeneity (observed or unobserved) diminish the inefficiency estimates they end up with very different rank orders of firms. This is of course very important information to the regulatory bodies.

The firm-specific inefficiency scores based on the TFE model are very close to each other and if we look at the variance of the frontier in this model we notice it to be rather big which shows that the model does not produce robust estimates for the frontier. This can be due to the rather short panel or insufficient number of observations. According to BIC criteria the model that combines the characteristics of unobserved and observed heterogeneity fits the data best and this points to the importance of taking unobserved heterogeneity into account.

References

1. Dalen DM, Gomez-Lobo A (2003) Yardsticks on the road: regulatory contracts and cost efficiency in the Norwegian bus industry. Transportation 30:371–386
2. Farsi M, Filippini M (2004) Regulation and measuring cost efficiency with panel data models application to electricity distribution utilities. Rev Ind Organiz 25:1–19

3. Farsi M, Filippini M, Greene W (2005) Efficiency measurement in network industries: application to the swiss railway companies. J Regul Econ 28:69–90
4. Farsi M, Filippini M, Greene W (2006) Application of panel data models in benchmarking analysis of the electricity distribution sector. Ann Public Cooperative Econ 77:271–290
5. Farsi M, Filippini M, Kuenzle M (2006) Cost efficiency in regional bus companies: an application of alternative stochastic frontier models. J Trans Econ Policy 40:95–118
6. Farsi M, Filippini M, Kuenzle M (2007) Cost efficiency in the swiss gas distribution sector. Energy Econ 29:64–78
7. Farsi M, Filippini M (2009) An analysis of cost efficiency in swiss multi-utilities. Energy Econ 31:306–315
8. Greene W (2004) Distinguishing between heterogeneity and inefficiency: stochastic frontier analysis of the World Health Organization's panel data on national health care systems. Health Econ 30:959–980
9. Greene W (2005) Reconsidering heterogeneity in panel data estimators of the stochastic frontier model. J Econometrics 126:269–303
10. Greene W (2005) Fixed and random effects in stochastic frontier models. J Prod Anal 23:7–32
11. Jha R, Singh SK (2001) Small is efficient: a frontier approach to cost inefficiencies in Indian state road transport undertakings. Int J Trans Econ 18:95–114
12. Kopsakangas-Savolainen M, Svento R (2008) Estimation of cost-effectiveness of the Finnish electricity distribution utilities. Energy Econ 30:212–229
13. Kopsakangas-Savolainen M, Svento R (2011) Observed and unobserved heterogeneity in stochastic frontier models: an application to the electricity distribution industry. Energy Econ 33:304–310
14. Kumbhakar SC, Hjalmarsson L (1993) Technical efficiency and technical progress in Swedish dairy farms. In: Fried HO, Lovell CAK, Schmidt SS (eds) The measurement of productive efficiency: techniques and applications. Oxford University Press, Oxford
15. Kumbhakar S, Lovell CAK (2000) Stochastic frontier analysis. Cambridge University Press, Cambridge
16. Pitt M, Lee L (1981) The Measurement and sources of technical inefficiency in Indonesian weaving industry. J Dev Econ 9:43–64
17. Schmidt P, Sickles RE (1984) Production frontiers and panel data. J Bus Econ Stat 2:367–374

Chapter 9
Regulating Electricity Distribution Utilities

9.1 Background

Although electricity generation and retail have been liberalized and opened up to competition in many countries significant portions of the total electricity supply system—distribution and transmission—continue to be regulated legal natural monopolies. In this chapter we concentrate on regulation and its practical applications.

It is generally known that fully informed regulators do not exist. In most cases the regulated firm has more information about its costs and other factors than the regulator. Accordingly, the regulated firm may use its information advantage strategically in the regulatory process to increase its profits or to pursue other managerial goals, to the disadvantage of consumers. Many regulatory agencies have put much effort into reducing this information asymmetry. Theoretical research on regulation (especially incentive regulation) has also evolved and it has provided new information to regulators (see, e.g., [1, 2, 11–13]). Because the regulator has less information than the firms the regulated firms have strategic advantage. Generally, any firm would like to convince the regulator that it is a "higher cost" firm than it actually is. By behaving like this the firm believes that the regulator sets higher prices (which increases firms' profits and transfers welfare from consumers to the regulated firms).

When a social-welfare-maximizing regulator tries to distinguish between firms with high cost endowments and firms with low cost endowments it faces an *adverse selection* problem. One possible solution to this problem is to use the firm's ex post realized costs to set regulated prices. This means that the regulator uses some form of "Cost of Service" [or Rate of Return (ROR)] regulation. However, when the regulator solves the adverse selection problem in this way it leads to another problem, namely to the *moral hazard* problem. This is because the loss of the opportunity for the firm to earn extra profits reduces managerial effort and consequently less managerial effort increases the firm's realized costs. This

M. Kopsakangas-Savolainen and R. Svento, *Modern Energy Markets*,
Green Energy and Technology, DOI: 10.1007/978-1-4471-2972-1_9,

leads to the situation, where regulation possibly increases the costs above their efficient levels.

The moral hazard problem may be solved by some form of incentive regulation (e.g., Price Cap regulation, Revenue Cap regulation or Yardstick Competition), but then the costs of adverse selection are incurred. Price Cap regulation is, however, very weak at rent extraction for the benefit of consumers and society and it potentially leaves a lot of rent to the firm. The task of the regulator, then, is to find such a regulation mechanism that takes the social costs of adverse selection and moral hazard into account. One of these methods is the so-called Menu of Contracts regulation, where for each firm a menu of cost-contingent contracts is offered and the firm can choose the contract which it prefers among the menu (see [11]).

The main contribution of this chapter comes from comparing welfare effects of different regulation schemes in electricity distribution utilities.[1] The four regulation schemes which we compare are Fixed Price regulation, Cost of Service regulation, Menu of Cost-Contingent Contracts, and Simple Menu of Contracts. In our calculations we utilize the estimated benchmarking information of firm-specific efficient costs. The firm-specific cost information is obtained by using various models of Stochastic Frontier Analysis (SFA). Stochastic frontier methodology and the essential literature is presented earlier in Chaps. 7 and 8. We have used inefficiency results of four model specifications presented earlier in Chap. 8 in this book. Our benchmark model is basic RE model (see Eq. 8.9). In addition we have used three modifications where both observed and unobserved heterogeneity has been taken into account such that they are not mixed into inefficiency measures. These modifications are called REH, TRE, and TFE and they are presented in Eqs. 8.10–8.12. For the more detailed information on data, stochastic frontier parameters and on inefficiency results we refer reader to see Tables 7.1, 8.2, and 8.3.

By using the information contained in the SFA results, the cost information asymmetry can be reduced and the regulator can evaluate the firm-specific potential for cost reduction if some form of incentive regulation is implemented. Using this information it is possible to calculate the social welfare effects of different regulation schemes.

Our basic result is that total welfare can be improved if we move from the Cost of Service regulation scheme to the Menu of Contracts regulation, Simple Menu of Contracts, or to the Fixed Price regulation. There is, however, a significant difference among regulation regimes in how this improved welfare is distributed between consumers and producers.

[1] See Pint [16] for comparing the welfare effects of Price-Cap and ROR regulation in a stochastic-cost model.

9.2 Different Regulation Models

9.2.1 Rate of Return/Cost of Service Regulation

The ROR or Cost of Service regulation is the traditional approach to regulate monopolies. In the purest form of this regulation the ROR is fixed to the costs. This means that the utility does not face the risk connected to input price changes or other risks related to the costs. There are naturally both advantages and disadvantages connected to each regulatory method. The advantages of ROR regulation are its relatively simple practical applicability, the possibility to use second-best (e.g. Ramsey-Boiteux) or nonlinear prices, the use of deliberate cross-subsidization,[2] and finally the rate hearings provide an opportunity for customers to express their views (see, e.g., [14]). It has also been argued that an additional positive feature of this regulation method is better certainty of long-term investments. Disadvantages connected to this regulation scheme (just to state the most obvious ones) are that it does not give incentives to produce efficiently and if the allowed ROR on capital is higher than the cost of capital an input bias (called the Averch–Johnson effect) follows. ROR regulation also typically entails high administrative costs (due to time-consuming hearings and requirement of considerable knowledge about the firm's costs). One further difficulty is in determining the "right" level of allowed ROR, which has been found to be problematic. On the other hand it is not possible for the firms to gather excessive profits or to incur big losses. Formally the ROR regulation[3] for firm i can be written as:

$$p_{i,t} = (1 + r)c_{i,t-1}, \tag{9.1}$$

where p is the allowed price for firm i at period t, r is the allowed ROR, and c is realized costs at period $t - 1$.

To summarize, the main reservation against this approach is that it clearly does not provide incentives for cost savings and efficiency improvements. It may also easily lead to overinvestment.

9.2.2 Price Cap Regulation

The Price Cap regulation has perhaps been the most significant alternative to the ROR regulation method in utility regulation. The Price Cap method was first proposed by Littlechild [15] and it has since been adopted in the regulation of many industries (including telecommunication, gas distribution, water distribution,

---

[2] E.g. local telephone services at low rates subsidized by long-distance services.

[3] When the ROR regulation is determined this way it is identical to the pure cost of service regulation where firm is allowed to add some fixed percent above its costs.

airline industry, railway industry, and electricity) in the UK and other countries as well.[4]

In its purest form either the price or the price path is fixed. This means that the utility faces the full risk connected to the input prices and demand. On the other hand, the utility has full incentive to reduce costs because it can keep all the benefits from cost reduction. This has been seen as the main advantage of this regulation scheme. The firm-specific potential for cost reduction can be evaluated by using, e.g., SFA-based inefficiency results. The other advantages are relatively small administrative costs, nonexistence of input bias (A–J effect), and the fact that price ceilings on monopoly services prevent predatory pricings. Because of the differentiation of the price from costs and the fact that in pure price regulation the price does not react to any exogenous factors, it is possible for the firm to gather excess profits or also to incur big financial losses. As in any regulatory scheme there exist also some disadvantages in this method. The disadvantages include the uncertainty about the service quality, the fact that potential benefits are ruled out (e.g. Ramsey-Boiteux prices), implementation of the price cap can be difficult, and there may be a greater possibility for capturing the regulatory process by the firm (see, e.g., [14]).

The Price Cap regulation essentially decouples the profits of the regulated utility from its costs by setting a price ceiling. This method is also referred to as the "RPI–X" model. In this model the price cap for each year is set based on the retail price index (RPI) and an efficiency factor X. Hence prices remain fixed for the rate period and the utility is allowed to keep the achieved cost savings. Formally the price ceiling for firm i is set according to the following equation:

$$p_{i,t} = p_{i,t-1}{}^{*}(\mathrm{RPI} - \mathrm{X}_i) +/- \mathrm{Z}_i \tag{9.2}$$

According to the equation the price ceiling p_t for each year is calculated based on the previous year's price ceiling p_{t-1} adjusted by RPI minus the efficiency factor X. The efficiency factor X is set by the regulator. In practice the price ceiling may be adjusted using a correction factor Z. This correction factor accounts for the effect of exogenous extraordinary events affecting the utility's costs.

The potential problems associated with the practical implementation of Price Cap regulation are connected to: the price review procedure, the commitment of the regulating authority to quality, and the rules how to determine the X parameter (see, e.g., [8], [9]).

9.2.3 *The Optimal Incentive Scheme*

It has been argued (see, e.g., [9, 11, 12]) that perhaps the optimal regulatory mechanism will lie somewhere between these two extremes, i.e., between the pure

---

[4] For the description and discussion of price cap and rate-of-return regulation see also Armstrong and Sappington [3], Joskow [8] and Liston [14].

Cost of Service (or ROR) and pure price regulation. The regulation model will then have the form of either a profit sharing contract or a sliding scale mechanism (price that the regulated firm can charge is partially responsive to changes in realized costs and partially fixed ex ante). It has also been argued that more generally, by offering a menu of cost-contingent regulatory contracts with different cost sharing provisions, the regulator can do even better than if it offers only a single profit sharing contract [12]. The basic idea of the optimal incentive scheme is to make it profitable for a firm with low cost opportunities to choose a relatively high powered incentive scheme (e.g., price or revenue cap regulation) and a firm with high cost opportunities a relatively low powered scheme (e.g., ROR or Cost of Service regulation).

In the Laffont–Tirole model [11] the firm chooses output and effort, and after the costs are realized, the planner rewards the firm according to the two observables, output and costs. Equivalently, the planner can ask the firm to reveal its true productivity parameter. Laffont and Tirole show that it is possible to construct such an incentive scheme that induces the firm to tell the truth and that the level of effort is voluntarily optimally chosen by the firm. The incentive scheme is linear in costs and can be written as $T(\beta, c) = s^*(\beta) + K(\beta)[c^*(\beta) - c]$, where T is net transfer to the firm, s* is ex ante reward, c* is ex ante cost, c is realized costs, β is the productivity parameter and $K^*(\beta) = \frac{\psi'[e^*(\beta)]}{q^*(\beta)}$, where q* is optimal output, e* is optimal effort level, and ψ' is marginal disutility of effort. Hence the reward depends on the announcement of the β and the ex post costs.

In practice the optimal allocation can be implemented by asking the firm to announce the expected average costs and by making the transfer depend on the expected and realized average costs. The ex ante reward and the slope of the ex post bonus scheme decreases with the announced cost. Chu and Sappington [6] consider a straightforward extension of Laffont and Tirole's model which allows the supplier to be able to reduce production more easily when cost are initially high than when they are initially low. Chu and Sappington also show how this extended model admits many of the forms of optimal contracts that prevail in practice.

The theory of a menu of cost-contingent regulatory contracts with different cost sharing provisions has two related problems when the practical implementation comes to question. These problems are first the fact that the economic logic and the underlying mathematics involved in calculating the optimal menu are quite complex, and second the issue that the principal must be able to specify the agent's entire disutility of effort function in order to calculate the optimal menu. Consequently the model has not been widely used in practice or even in empirical applications of the theory.

9.2.4 Simple Menu of Contracts

Rogerson [18] shows that dramatically simpler menus (than the Optimal Incentive Scheme) which are easy to understand and calculate and which have lower informational requirements can capture a substantial share of the gains achievable by the fully optimal complex menu.[5] The problem for the principal (or regulator) is to find such a menu of contracts that minimizes his expected payment to the agent, subject to the constraint that all types of the agents accept a contract and produce the good. Rogerson uses the name "fixed price cost-reimbursement" (FPCR) for a simple contract menu. Pure cost-reimbursement contracting corresponds to ROR regulation and fixed price contracting corresponds to Price Cap regulation. In his paper Rogerson shows that there is a unique optimal FPCR menu which solves the principal's cost minimization problem. In order to use this simple menu in real contracting situations, a principal would need to have information on the cumulative distribution and density of costs if a cost-reimbursement contract is used and the size of the efficiency gain that he believes would be induced by fixed price contracting. Again, the information needed in evaluating the efficiency gain can be provided by the frontier estimation results.

The procedure to determine the FPCR menu is the following: first, the principal offers the agent a menu of contracts specifying price as a function of costs, and then the agent decides which contract to accept. A theoretical lower bound on the principal's expected price is the price equal to procurement cost in the case he had full information regarding the type of the agent. The principal would minimize his costs by offering to pay an agent of type x a fixed price equal to $x - k$, where k denotes the resulting surplus. If the agent accepts this contract, he reduces costs to the efficient level (or first-best level) and earns zero profits. The upper bound on the principal's expected price is the price which guarantees that agent will always produce the good, i.e., the principal will offer the agent a cost-reimbursement contract in which the principal promises to pay the agent a price equal to the measured cost of production. Under such a contract, the agent has no incentive to reduce costs and again earns zero profits. The problem of the principal is to find the "cut-off type" θ so that his expected payment to the agent is minimized. The "cut-off type" is the highest type willing to accept the Fixed Price contract. Rogerson [18] shows that if there exists an $x^* \in [x_{min}, x_{max}]$, which solves $F(x^*) = kf(x^*)$ then $\theta = x^*$. However, if $F(x) < kf(x)$ for every $x \in [x_{min}, x_{max}]$ then $\theta = x_{max}$. F(x) is the distribution function of x and f(x) is the density function of x. We show later how to apply this model in practice (see also [10]).

[5] See also Bower [5], Gasmi et al. [7], Reichelstein [17], and Sappington and Weisman [19] for studying the performance of simple mechanism in Laffont-Tirole type framework.

9.3 Regulation and Welfare

9.3.1 Welfare Calculations

Next we combine our empirical SFA cost-inefficiency information with different regulation schemes and calculate changes in total social welfare related to each model specification and regulation scheme. In these calculations we use Cost of Service regulation as the benchmark.

The inefficiencies are naturally endogenous to the regulation scheme in use. In this sense our results are short-term efficiency gains resulting from immediate change of the scheme in use. The endogenous long-term effects escape our results because there is no data to be used for this test. However, the endogenous long-term effects can safely be expected not to reduce the short-term welfare gains calculated here. This view is based on the intuition that, if a new regulation scheme provides incentive to improve cost efficiency in the short term it is unlikely that the same incentive mechanism reduces efficiency on the long run. Actually, efficiency may further increase in the long run as the firm and its management has the opportunity to learn from the past and consequently make better long-run efficient decisions regarding investments, management, personnel policy, etc.

The overall social welfare change is calculated as the sum of the change in producer and consumer surpluses. These changes in consumer and producer surpluses can be specified as follows. The change in consumer surplus can be written as the line integral

$$\Delta \mathrm{CS}_i = \int_{P_{\mathrm{C}i}}^{P_{\mathrm{N}i}} \mathrm{D}^{-1}(Q_i)\mathrm{d}Q_i, \tag{9.3}$$

where $\Delta\mathrm{CS}_i$ is the change in consumer surplus for customers of firm i, $P_{\mathrm{C}i}$ is the price of firm i under Cost of Service regulation, $P_{\mathrm{N}i}$ is the new price of firm i under either the Fixed Price, Menu of Contracts[6] or Simple Menu of Contracts regulation, D^{-1} is the constant elasticity inverse demand function with the elasticity value[7] of -0.35 and Q_i is distributed total energy of firm i. The corresponding change in producer i's surplus is:

$$\Delta \mathrm{PS}_i = (P_{\mathrm{N}i}Q_{\mathrm{N}i} - C(Q_{\mathrm{N}i})) - (P_{\mathrm{C}i}Q_{\mathrm{C}i} - C(Q_{\mathrm{C}i})), \tag{9.4}$$

---

[6] We use the names "menu of contracts" and "fixed price" regulation for our applications of the incentive contract scheme and price cap regulation respectively.

[7] According to Törmä [20] the price elasticity of electricity demand in Finland was approximately −0.35. Also the results by Andersson and Damsgaard [4] and Willner [21] support the assumption of inelastic demand.

where the new volume Q_{N} is the supply and demand equilibrating quantity at the new price[8] P_{N} and $C(Q_{\mathrm{C}i})$ are the costs resulting from Cost of Service regulation and $C(Q_{\mathrm{N}i})$ is the new costs resulting from the alternative regulation scheme and based on our efficiency estimation results.

Therefore, the change of total surplus is

$$\sum_{i=1}^{76}(\Delta \mathrm{CS}_i + \Delta \mathrm{PS}_i) =$$
$$\sum_{i=1}^{76}\left[\int_{P_{\mathrm{C}}}^{P_{\mathrm{N}}} \mathrm{D}^{-1}(Q_i)\mathrm{d}Q_i + [(P_{\mathrm{N}i}Q_{\mathrm{N}i} - C(Q_{\mathrm{N}i})) - (P_{\mathrm{C}i}Q_{\mathrm{C}i} - C(Q_{\mathrm{C}i}))]\right] \tag{9.5}$$

9.3.2 The Regulation Application

The range of regulatory options in our welfare calculations can be illustrated as follows. Consider a regulatory process in which the firm's allowed price P is determined based on a component of the costs of the firm with the highest efficient costs, C*, and on a component that is based on the firm's own realized costs C_i. The efficient cost of the highest type C* is obtained from our SFA estimations. Then the allowed price is determined according to the following equation[9]:

$$P_{\mathrm{N}i} = a\mathrm{C}^* + (1 - a)C_i, \tag{9.6}$$

where a is the sharing parameter that defines the responsiveness of the firm's allowed price to the realized costs. In the case of Fixed Price contract (Price or Price Cap regulation) $a = 1$, whereas in the case of pure Cost of Service (or ROR) regulation, assuming that the regulator can observe the firm's expenditures but not evaluate their efficiency, $a = 0$. The profit sharing contract (or Menu of Contracts in our case) emerges with $0 < a < 1$.

Laffont–Tirole [11, 12] show that it is socially optimal for the regulator to offer a menu of contracts with different combinations of responsiveness parameters. This should drive the firms with low true cost opportunities to choose a high powered scheme (a close to 1) and consequently firms whose true (efficient) cost are high to choose a lower powered incentive scheme (a close to zero). In calculating the welfare effects we assume that each firm chooses the efficiency level which maximizes its profits.

---

[8] This assumption can be made because real-time demand and supply must always equal each other in electricity distribution.

[9] In the Eq. 9.6 the efficient cost of the highest type C* is calculated as a 6-year average based on the yearly estimation results and C_i is the average value of the firms realized cost.

9.3.2.1 Fixed Price Regulation

In fixed price regulation the value of parameter a is equal to 1 for each firm and the C* is the efficient cost of the highest type firm obtained by utilizing the SFA results: $P_{Ni} = C^*$. Now the profit maximizing strategy for each firm is to produce at efficient costs and we have evaluated these efficient costs from our SFA estimations. In the Cost of Service regulation (which acts as benchmark model in our welfare calculations) the value of parameter a is equal to zero for each firm and consequently firms produce at zero profits.

9.3.2.2 Menu of Contracts regulation

The game behind the Menu of Contracts regulation scheme in our application is as follows: first the regulator announces the regulation rule, i.e., the rules according to which the value of parameter a is determined and then the firm decides what efficiency level to choose (and we assume that it chooses the efficiency level which maximizes its profits). The value of parameter a is defined as follows. The regulator orders the firms in descending order according to their firm-specific efficiency scores obtained as a result of SFA estimations. The most efficient firm, i.e. the firm which has the relatively smallest inefficiency score, is referred as $\mathrm{eff}^{\max}$. The most inefficient firm based on its inefficiency score is referred as $\mathrm{eff}^{\min}$. The value of parameter a for each firm i in this application of Menu of Contracts regulation is set following

$$a_i = 1 - \frac{eff^i - eff^{\max}}{eff^{\min} - eff^{\max}}, \quad i = 1, \ldots, 76. \tag{9.7}$$

Using Eq. 9.7 we in fact rescale our efficiency results so that they vary from zero to one: $0 \le \mathrm{eff}^{\,i} \le 1$. According to this announced rule for parameter a, for the most efficient firm the value of a is equal to 1, and thus this firm is allowed to set the price such that it equals the efficient cost of the highest cost type. For the most inefficient firm the value of parameter a is equal to zero and consequently the price is equal to its realized costs and the firm earns zero profits. For firms between the most efficient and most inefficient firms, the value of parameter a is bigger for the more efficient firms than those which are more inefficient. As a result of this rule the profit maximizing strategy for each firm is to produce at efficient costs.

9.3.2.3 Simple Menu of Contracts Regulation

The price determined by the Simple Menu of Contracts is calculated according to the rule described in Sect. 9.2.4. According to that rule the objective of the regulator is to choose the so-called “cut-off type” such that the costs of the regulator are minimized if it should cover the procurement costs. The “cut-off type” is the

Table 9.1 Change in average yearly welfare (TS = PS + CS), cost of service regulation[a] as benchmark, million €

SFA model	Fixed price ΔTS	ΔPS	ΔCS	Menu of contracts ΔTS	ΔPS	ΔCS	Simple menu of contracts ΔTS	ΔPS	ΔCS
RE	177.8	240.9	−63.1	194.4	150.0	44.4	144.43	70.6	73.9
REH	49.6	234.2	−184.5	61.5	184.1	−122.6	25.9	10.4	15.6
TRE	5.6	239.7	−234.1	25.8	163.6	−137.8	6.4	4.5	1.9
TFE	8.3	235.5	−227.2	14.7	207.4	−192.7	3.6	1.4	2.2

[a] We have used 4% for the r in cost of service regulation (see Eq. 9.1)

highest type willing to accept the Price Cap contract. The cut-off type (θ) is found by solving the equation F(x*) = kf(x*) (then θ = x*), where x is the agent's type, F(x) is the distribution function of x and f(x) is the density function of x. In the Simple Menu of Contracts the firm can choose between either Price Cap or Cost of Service regulation. Now each firm more or as efficient as the cut-off type firm maximizes its profits by choosing the Price Cap contract (parameter *a* equal to 1) and the best strategy for rest of the firms is to choose Cost of Service regulation (parameter *a* equal to zero).

9.3.3 The Welfare Results

Based on these regulation schemes we are able to calculate the total welfare changes using our firm-specific results on efficiency improvement potentials. The results for all four SFA model types (see Eqs. 8.9–8.12) are presented in Table 9.1.

Changing the regulation scheme from Cost of Service to whatever other regulation regime presented above results in welfare improvements. The change in welfare is quite significant at least in the case of RE- and REH-based SFA models. If the potential for efficiency improvements are evaluated according to the RE model the resulting welfare change is 33.1% from the total value of distribution in Fixed Price regulation, 36.2% in the Menu of Contracts and 26.9% if the new regulation is based on the Simple Menu of Contracts. The corresponding percentage changes in welfare based on the REH model are 9.2% for the Fixed Price, 11.5% for the Menu of Contracts, and 4.8% for the Simple Menu of Contracts regulation schemes. Notable is, however, that if the potential to improve efficiency is based on either the TRE or the TFE model, the resulting changes in welfare are smaller. The TFE model fails to estimate the frontier distribution robustly (see Table 8.3) for the variance of the frontier error term) and is thus not very interesting here. Interesting is that randomized modeling of heterogeneity (TRE) reduces the potential welfare gains substantially. Part of this can be explained by the fact that the firms in question are local monopolies and it is possible that they do not put maximum effort into achieving efficiency. Consequently, part of the time invariant inefficiency (now assumed to be due to firm-specific unobserved heterogeneity) may be due to inefficient management and hence the model TRE may underestimate the inefficiency scores. This result stresses the need to model hetereogeity correctly in SFA models.

Another observation from the welfare results is that there is a clear difference in how various regulation schemes divide welfare between producers and consumers. In the case of Menu of Contracts both the producer surplus and consumer surplus increase if the efficient levels of the firms' costs are determined by using the values of the random effects (RE) model. If the possibilities of the efficiency improvements are determined according to the REH, TRE, or TFE models producer surplus clearly increases but the consumer surplus decreases. This is due to the fact that the efficiency improvement possibilities according to the REH, TRE, and TFE models are smaller than according to the model RE and consequently the efficient cost of the highest type C* (see Eq. 9.6) is quite high. This raises the average level of the allowed price and hence transfers the welfare from the consumers to the producers. This result is even clearer in the case of Fixed Price regulation, where all firms are allowed to set the price equal to the efficient cost of the highest type (parameter a is equal to 1 for every firm). The only regulation scheme which improves both producer and consumer welfare regardless of the model used in efficiency estimations is the Simple Menu of Contracts. However, the overall welfare improvement is smaller than that resulting from the Fixed Price regulation or Menu of Contracts regulation. If the regulator is only interested in maximizing total welfare it should choose Menu of Contracts regulation. In the case where the regulator is more concerned about consumer welfare and wishes to see lower overall prices, she should set the regulation according to the Simple Menu of Contracts. In many countries, during the restructuring process of electricity industry, the main role of the electricity distribution network is designed to be a neutral market place for competitive parties of the industry. Based on this view, the regulator might be more interested in consumer welfare than providing possibilites for higher profits to the local electricity distribution monopolies. Regardless of the target of the regulator it is clear that the results support the theory, and changing the regulation method of network prices from traditional Cost of Service regulation it is possible to considerably improve social welfare.

The purpose of this chapter was to analyze whether it is possible to improve social welfare by changing the regulation scheme of electricity distribution and how the regulator can utilize information on firm-specific cost efficiency obtained by SFA models in the regulatory process. A great deal of theoretical research has been conducted concerning different regulation methods, but the connections of regulation theory to the real regulatory processes have been seen as problematic. Here we combine the theory of different regulation schemes to the firm-specific cost information of electricity distribution utilities obtained by using various Stochastic Frontier models. According to our results—which are consistent with the theory—Fixed Price regulation solves the problem of moral hazard and welfare improves if we move from Cost of Service regulation to the Fixed Price regulation. However, in Fixed Price regulation the problem of adverse selection remains unsolved. According to theory the Menu of Contracts regulation should solve both the moral hazard and adverse selection problems. Our empirical results support this. Welfare can be improved by changing from the pure cost-based regulation to the Menu of Contracts regulation. Notable is that welfare increases in all model specifications. However, there are significant differences among regulation schemes in how improved welfare is distributed to consumers and producers. Of the regulation schemes studied here the Fixed Price and the Menu of Contracts clearly emphasize the producer surplus benefits. Accordingly, if the regulator wants to give maximum incentive for cost reduction she should choose one of these schemes over Cost of Service regulation. The only regulation scheme which improves both producer and consumer welfare regardless of the model used in efficiency estimations is the Simple Menu of Contracts.

References

1. Armstrong M, Cowan S, Vickers J (1994) Regulatory reform: economic analysis and British experience. MIT Press, Cambridge
2. Armstrong M, Sappington DEM (2004) Toward a synthesis of models of regulatory policy design with limited information. J Regul Econ 26:5–21
3. Armstrong M, Sappington DEM (2007) Recent developments in the theory of regulation. In: Armstrong M, Porter R (eds) Handbook of industrial organization, vol 3, pp 1557–1700
4. Andersson B, Damsgaard N (1999) Residential electricity use—demand estimations using Swedish micro data. In: Paper presented at the 22nd IAEE annual international conference, Rome, 9–12 June
5. Bower AG (1993) Procurement policy and contracting efficiency. Int Econ Rev 34:873–901
6. Chu LY, Sappington DEM (2009) Procurement contracts: theory vs practice. Int J Ind Organ 27:51–59
7. Gasmi F, Laffont J–J, Sharkey WW (1999) Empirical evaluation of regulatory regimes in local telecommunications markets. J Econ Manage Strategy 8:61–94
8. Jamasb T, Pollitt M (2001) Benchmarking and regulation: international electricity experience. Util Policy 9:107–130
9. Joskow P (2008) Incentive regulation and its application to electricity networks. Rev Network Econ 7:547–560
10. Kopsakangas-Savolainen M, Svento R (2010) Comparing welfare effects of different regulation schemes: an application to the electricity distribution industry. Energy Policy 38:7370–7399
11. Laffont J–J, Tirole J (1986) Using cost observation to regulate firms. J Political Econ 94:614–641
12. Laffont J–J, Tirole J (1993) A theory of incentives in regulation and procurement. MIT Press, Cambridge
13. Laffont J–J (1994) The new economics of regulation ten years after. Econometrica 62:507–537
14. Liston C (1993) Price-cap versus rate-of-return regulation. J Regul Econ 5:25–48
15. Littlechild SC (1983) Regulation of British telecommunications' profitability. Her majesty's stationery office. Report to the Secretary of State, Department of Industry, London
16. Pint EM (1992) Price-cap versus rate-of-return regulation in a stochastic-cost model. RAND J Econ 23:564–578

17. Reichelstein S (1992) Constructing incentive schemes for government contracts. An application of agency theory. Acc Rev 67:712–731
18. Rogerson WP (2003) Simple menus of contracts in cost-based procurement and regulation. Am Econ Rev 93:919–926
19. Sappington DEM, Weisman DL (1996) Designing incentive regulations for the telecommunications industry. AEI Press, Washington
20. Törmä H (1985) Industrial demand for energy in Finland 1960–1982. Discussion paper no. 9, University of Jyväskylä, Finland
21. Willner JA (1996) Comment on bradburd: privatisation of natural monopolies. Rev Ind Organ 11:869–882

Chapter 10
The Future of Electricity Markets

10.1 Background

The challenging developments of the new energy system that have been analyzed in this book necessitate a change from the traditional "dumb grid" to an intelligent and adaptive "smart grid." This change in both transmission and distribution grids is well under way in many countries and large-scale effects of this transformation can be expected in the near future. Here we summarize the basic economic features of this change.

Smart Grids have been defined in many ways. Independently of the used definition there are some basic technological developments that are a necessary condition for an intelligent grid. These technological developments are designed to improve the performance of transmission and distribution systems by:

- "Installing sensors that can detect system conditions that indicate failures that either have occurred or will occur in the near future;
- Incorporating fast-acting microprocessors that can quickly detect fault conditions and take action to anticipate failures and reconfigure circuit supply routes or restore service as quickly as possible to customers that can be served by alternative supply lines;
- Reconfiguring radial circuits adding normally open points with automatic switches that can be closed to restore service to customers surrounding isolated faults automatically;
- Adding voltage regulation and capacitance down-stream of substation transformers reduce line losses—thus improving energy efficiency;
- Installing AMI (Automatic Metering Infrastructure) meters that provide a wide range of benefits including:
  - Reduced cost of meter reading
  - Improved ability to detect outages and restore service quickly after outages
  - Improved theft detection

M. Kopsakangas-Savolainen and R. Svento, *Modern Energy Markets*,
Green Energy and Technology, DOI: 10.1007/978-1-4471-2972-1_10,

- Improved access for customers to information about timing and magnitude of electricity consumption" [9].

Automatic Metering Infrastructure offers also many possibilities for creation of new incentive base mechanisms in order to affect consumers' behavior. Through AMI new type of electricity contracts between the electricity utility and final consumer can be exploited. One of these kinds of incentive-based mechanisms is the Real-Time Price-based contract already analyzed earlier in this book. Through new incentive-based mechanisms considerable efficiency improvements and welfare gains can potentially be received.

Once these technological improvements are in place the grid can be called smart and defined, e.g., like the DOE (U.S. Department of Energy): "A Smart Grid uses digital technology to improve reliability, security and efficiency of the electric system: from large generation, through the delivery systems to electricity consumers and a growing number of distributed generation and storage resources" [11]. A crucial amendment in this definition to the technological enablers is the possibility to integrate distributed resources to the system. In fact, Holt et al. [4, p. 2] refer to smart grid as "an electric transmission and distribution system where two-way communication exists between the source and the sink for the electricity." This two-way communication makes it possible for the economic agents connected to the smart grid behave both as suppliers and demanders of electricity thus changing the whole economic energy institution from a one-sided to a two-sided market place.

This market place logic is another way to see the elements and structure of the smart grid (see, e.g., [5]). In the supply side we can envision new virtual power plants to come into the world to enhance the controllability of the system by diversification. The basic idea of a virtual power plant is to connect distributed and renewable power generating facilities via modern ICT. A central control entity of the virtual power plant continuously monitors the generation data and has the power to switch individual generators in and out of the system.

The core effect of a smart grid on the demand side is to make consumption adaptive to system changes. We have already analyzed Real-Time Pricing as a core instrument in bringing this demand side adaptability into everyday operations of agents. Obviously modern ICT plays a crucial role also here. This led to DOE further to explain that "the information networks that are transforming our economy in other areas are also being applied to applications for dynamic optimization of electric system operations, maintenance and planning" [12].

10.2 Network Development and Potential Benefits

It is obvious that the change to smart grids brings various benefits both to individual agents and to the society as a whole. These benefits have been classified by Sullivan and Schellenberg [9, 10] as reliability benefits, economic benefits, and societal benefits. Reliability benefits can be identified as the reduced costs to utilities and customers resulting from service interruptions and power quality disturbances. These benefits and related costs have been analyzed quite extensively (e.g., Sullivan and Schellenberg [9, 10]). This research activity is natural since utilities are basically the sole carriers of smart grid construction costs and thus need information on the investment returns.

The economic benefits arise from optimized use of the electric system in the sense that peak load demand can be reduced as we have already shown. Also improved asset utilization in the sense of optimized generator operation, better planning and timing of investments, and reduced service costs can be expected to bring economic benefits.

According to Sullivan and Schellenberg the societal benefits come from environmental benefits related to optimized use of the electric system and from the possibility to integrate distributed energy resources devices to lower voltage distribution grids. These distributed energy resources devices break the pattern of economies of scale related to high-voltage transmission and distribution grids. According to our view especially environmental benefits can be even wider. Clear benefits can result through the change of consumers' behavior. Changing consumers' behavior requires new incentive-based mechanisms for which smart grid technology offers the required technical conditions. We have shown earlier in this book what kind of benefits can be reached through intelligent use of the pricing system.

Based on a broad survey of a large amount of electricity organizations from 11 European countries Donkelaar and Scheepers [3] and Donkelaar [2] reach a number of conclusions related to distributed generation. They see that investments in the technical options potentially improve the integration of distributed generation in several ways:

- "increased or optimized power production (distributed generation operator),
- access to markets for balancing and ancillary services (DG operator),
- reduced balancing costs (energy supplier),
- ability to construct a more exact E-program, and better comply with the E-program (energy supplier),
- improved power quality (distribution system operator),
- reduced operational and capital expenditures (distribution system operator)." [3, p.10]

The costs and benefits related to new technical solutions should be efficiency and justly allocated in order to receive maximum benefits for all parties. This also creates new challenges for the regulatory framework. One of these challenges is the question how to allocate the benefits between energy suppliers and distributed system operators in the absence of financial relationships. Another challenge raises from the allocation of indirect benefits which also have to be included in the new regulatory framework.

The analyses performed with the tool Donkelaar and Scheepers [3] develop show a number of benefits and costs that can be taken into account when parties involved in the electricity supply invest in new technical solutions and options to integrate distributed generation. They raise the need to quantify these benefits and costs identified and, of equal importance, the regulatory constraints that limit a "flexible" allocation of costs and benefits between distribution network actors.

10.3 Distributed Energy Resources as a Two-Sided Market

10.3.1 Theoretical Framework

The development and usage of smart grids open up new possibilities for decentralized energy production and use. Local markets can be expected to arise wherever local conditions for combined conventional power production together with good conditions for renewable energy sources like wind power and photovoltage power exist.

The basic economic question for such a local market is what is the local price structure and pricing logic in order to back up the existence of such markets? So how does a local aggregator platform (virtual power plant) set local energy prices in order to maximize the number of households and firms to participate in the local smart grid-based market? This setup can be modeled as a two-sided market (see, e.g., [8]). The basic electricity market with firms and households reacting to the offers given by utilities is not two-sided because the total volume of the market is not depending on the price structure under some aggregate restrictions related to the price level which is the usual definition of two-sidedness. The smart grid possibility of each agent potentially being a seller or/and buyer in the local market makes a big difference. It is obvious that the price structure inside the local market has an effect on the volume of the local market.

Let p^L be the local price of electricity, p^N be the national market price, and c be the cost of selling the electricity to the local broker. Also let N^B be the number of buyers in the local market and N^S the corresponding number of sellers. The utilities of buyers and sellers can then be written as:

$$\begin{aligned} U^B &= p^N - p^L\left(N^S, N^B\right) \\ U^S &= p^L\left(N^S, N^B\right) - c, \end{aligned} \tag{10.1}$$

where U^B denotes the utility of a buyer and U^S the utility of a seller and the local price function is assumed to be smooth and concave with respect to N^B and convex with respect to N^S. Assume that the network size affects the local price such that the larger the seller community, the lower shall the local price be and vice versa for the buyer community, i.e., the larger the buyer community, the higher the local price. Put in this way the network size is not a usual externality but note that the effect is asymmetric and so it creates a two-sided effect on this market. The interpretation of these utility functions is clear, the lower the local price with respect to the national price, the greater the buyer utility and the higher the local price with respect to the selling cost, the higher the seller utility.

An agent shall participate in the local market if her utility is non-negative. The participation restriction can thus be written as

$$\begin{aligned} U^B \geq 0 &\Rightarrow; \left[p^L\left(N^S, N^B\right) \leq p^N\right] \\ U^S \geq 0 &\Rightarrow; \left[p^L\left(N^S, N^B\right) \geq c\right]. \end{aligned} \tag{10.2}$$

Since each member of the local seller and buyer community can have both roles in the two-sided market, we can solve the numbers of buyers and sellers N^S and N^B as functions of p^N, p^L, and c

$$\begin{aligned} N^B &= n^B(p^N, p^L, c) \\ N^S &= n^S(p^N, p^L, c). \end{aligned} \tag{10.3}$$

The virtual power broker can differentiate between seller and buyer prices. Assuming that the national price and the local supply cost are exogeneous with respect to the membership volume we can simplify the endogeneous part of the local system into

$$\begin{aligned} N^B &= n^B(p^B, p^S) \\ N^S &= n^S(p^B, p^S), \end{aligned} \tag{10.4}$$

where p^B denotes the local demand price and p^S the local supply price. Now the local broker wants to maximize the size of the local network. Here we keep to the usual assumption in the literature that the volume can be expressed in a multiplicative form. In practice the exact volume is of course a sum of the numbers of buyers. The maximization problem of the broker is thus

$$\max\{V(p^B, p^S) = n^S(p^B, p^S)n^B(p^B, p^S) \quad \text{s.t. } p^N > p^B \geq p^S \geq c\}, \tag{10.5}$$

where $V(p^B, p^S)$ is the volume of the local network. Assuming that the first inequality in the restriction must always be effective for the local market to exist we can concentrate on the local price structure. The Lagrangian is

$$L = n^S(p^B, p^S)n^B(p^B, p^S) + \lambda(p^B - p^S) + \gamma(p^S - c). \tag{10.6}$$

The Kuhn-Tucker necessary conditions are

$$\begin{aligned} &\frac{\partial L}{\partial p^S} = \frac{\partial n^S}{\partial p^S} n^B + \frac{\partial n^B}{\partial p^S} n^S - \lambda + \gamma = 0 \\ &\frac{\partial L}{\partial p^B} = \frac{\partial n^S}{\partial p^B} n^B + \frac{\partial n^B}{\partial p^B} n^S + \lambda = 0 \\ &\frac{\partial L}{\partial \lambda} = p^B - p^S \geq 0, \quad \frac{\partial L}{\partial \gamma} = p^S - c \geq 0, \quad \lambda \geq 0, \quad \gamma \geq 0 \\ &\lambda \frac{\partial L}{\partial \lambda} = \lambda(p^B - p^S) = 0, \quad \gamma \frac{\partial L}{\partial \gamma} = \gamma(p^S - c) = 0. \end{aligned} \tag{10.7}$$

Assuming that the supply price condition $p^S > c$ is always binding, we can proceed with $\gamma = 0$. Then we are left with two possible solutions. Either the local broker is a non-profit cooperative and we can find the solution $p^B = p^S$ or the broker is a for-profit firm with $p^B > p^S$. Let us look at the first case. Assuming $p^B = p^S$ we have $\lambda > 0$ and the solution is of the form

$$\frac{\partial n^S}{\partial p^S} n^B + \frac{\partial n^B}{\partial p^S} n^S + \frac{\partial n^S}{\partial p^B} n^B + \frac{\partial n^B}{\partial p^B} n^S = 0. \tag{10.8}$$

Rewriting yields

$$\frac{\partial n^S/\partial p^B + \partial n^S/\partial p^S}{n^S} = -\left[\frac{\partial n^B/\partial p^B + \partial n^B/\partial p^S}{n^B}\right]. \tag{10.9}$$

The price structure is optimal from the broker's side when changes in both buyer and seller price have symmetric effects on the relative sizes of buyer and seller communities. This is a usual market equilibrium condition saying that prices can be increased as long as the changes in seller side are bigger than in the buyer side.

This result can be expressed also in elasticity terms. Remembering that here $p^S = p^B = p^L$ and multiplying and dividing (10.9) in both sides with p^L we have

$$e^S = -e^B, \tag{10.10}$$

where e^i, $i = S, B$ is the corresponding elasticity of volume with respect to the local price. This again means that the relative changes in supplier and buyer volumes must be equal (in absolute terms). If the relative increase in the number of sellers is greater than the relative decrease in the buyer community, the price can be increased. For the profit maximizing broker ($p^B > p^S$ and $\lambda = 0$) the volume maximizing price structure can be solved from

$$\frac{\partial n^S}{\partial p^S} n^B + \frac{\partial n^B}{\partial p^S} n^S = \frac{\partial n^S}{\partial p^B} n^B + \frac{\partial n^B}{\partial p^B} n^S. \tag{10.11}$$

Again restructuring and multiplying by $1/(n^B n^S)$ yields

$$\frac{\partial n^S/\partial p^S - \partial n^S/\partial p^B}{n^S} = \frac{\partial n^B/\partial p^B - \partial n^B/\partial p^S}{n^B}. \tag{10.12}$$

Now we see that increases in buyer and seller prices starting from an asymmetric standing change the relative sizes of both communities differently than in the symmetric case. Now the numerators can be either positive or negative. Again the price structure must be such that its change leads to symmetric changes in the numbers of sellers and buyers. If this is not the case the market cannot be in equilibrium.

Also this condition can be expressed in elasticity form. Now the solution can be expressed as

$$e^{BB} + e^{SB} = -\left(e^{SS} + e^{BS}\right), \tag{10.13}$$

where the first uppercase letter indicates the community and the second the price. Now the net effect of the buyer price change must in relative terms equal the corresponding effect of the seller price.

10.3.2 Discussion on Distributed Energy Sources

The management of this kind of a two-sided market necessitates totally new approaches and attitudes concerning the development of intelligent networks. A few proposals have, however, already been made into this direction. Overbeeke and Roberts [7] present a vision for 'Active Networks' as facilitators for the kind of distributed generation that we have modeled above. They foresee that passive distribution networks, as we know them, have to evolve gradually into actively managed networks. From their viewpoint it is both technically and economically the best way to facilitate distributed generation in a deregulated electricity market.

In the active networks vision, the principles of network management differ from the classical view of networks, being only one-way lanes for electricity transport from high-voltage to low-voltage grids. First of all, the network should not be considered as a power supply system but as a highway system that provides connectivity between points of supply and consumption. The "infinite network" as customers used to know it, no longer exists. A network interacts with its customers and is affected by whatever loads and generators are doing.

To change the network infrastructure based on these principles Overbeeke and Robers [7] proposed the concepts for structural solution. These concepts can be summarized to include the following views[1]; Interconnection of networks as opposed to dominantly radial networks meaning a switch from thinking one-directional to bi-directional flows; System includes local control areas ('cells') and consequently use automation to support relatively small control areas; System services are specified attributes of a connection—referring to the way in which system services provided by different units are charged to individual customers; Energy transport is not dependent on single part, so the vulnerability to component failures is significantly reduced; and preventing the domino effect (faults propagating over a very large area) by isolating faults so that the rest of the system can operate normally.

Overbeeke and Robers [7] introduced the concept of 'cells', which are in fact "local control areas." The cell concept does not have a large impact on the topology of the power network, the difference is the control hierarchy. Each cell will eventually have its own power control system, essentially computer-based, which manages the flow of power across the cell's boundaries. In the future this means that control systems of adjacent cells will negotiate in real time how much power will be transferred over their mutual interconnection.

The most obvious advantage of introducing Active Networks is that the changes proposed ask for virtually no physical reinforcement, meaning significant economic benefits. Those reinforcements are unavoidable if we are to accommodate larger amounts of distributed generation within a traditional system. The Active Networks vision has clearly economic advantages above traditional forms of upgrading. Firstly, this kind of system and hierarchy structure requires only a few

---

[1] The list of concepts is from Overbeeke and Roberst [7].

additional power lines. New lines are basically required on to provide interconnection between islands. Secondly, the already existing lines may be reinforced. This is applicable mostly to tapered circuits where local voltage control is uneconomical. Thirdly, this kind of system requires no new transformers and further interconnection improves security of supply and existing transformers can be operated to a higher percentage of the rated load. Fourthly, there is more switchgear and this increase options for inter- and disconnection. In order to secure that inter- and disconnection is more easy all switches have to be remotely operated. And lastly there is more control of investments by phased introduction of the system (see [7]).

Smart grids can be also described by the concept of Micro-Grids [3, pp. 29–30, 6]. Micro-Grids are small power systems that can operate independently of the bulk power system. They are composed of distributed energy production and energy-storage resources interconnected by a distribution system. They may operate in parallel with the bulk supply system during normal conditions and transform to islanded (stand-alone) operation during abnormal conditions such as an outage in the bulk supply or emergency. Micro-grids may also be created without connection to a bulk supply and operate full-time as an independent island.

The common question of how much penetration of distributed generation the grid can handle before stability problems result is not an issue with Micro-Grids because they are designed to satisfy their predetermined local load without creating any stability problems for the transmission system. Potential Micro-grid designs range in size from a single house operated independently up to a large substation-scale system that serves many feeders where total load may approach 100 MW.

Micro-Grids offer the potential for improvements in energy delivery, efficiency, reliability, power quality, and costs of operation as compared to traditional power systems. Micro-Grids can also help overcome constraints in the development of new transmission capacity that are beginning to impact the power industry.

One of the more interesting findings related to Micro-Grids research is that the use of uniformly distributed generation on Micro-Grids facilitates the ability to build distribution systems that do not need any high-voltage elements; they are entirely low-voltage. This low-voltage approach has potential for significant cost-savings and power quality/reliability improvements and can provide improved safety benefits as well.

A key motivation of Micro-Grids is the desire to move control of power reliability and quality closer to the point of end-use so that these properties can be optimized for the specific loads served.

The power grid can benefit from Micro-Grids by reducing congestion and other threats to system adequacy if they are deployed as interruptible or controlled loads that can be partially shed as necessary in response to changing grid conditions. Furthermore, Micro-Grids could provide the possibility to operate some or all of its end users at lower costs than would be possible on the traditional grid. The costs of delivered energy from the traditional power system includes losses, customer services, congestion, and other costs that together typically exceed the generation cost alone.

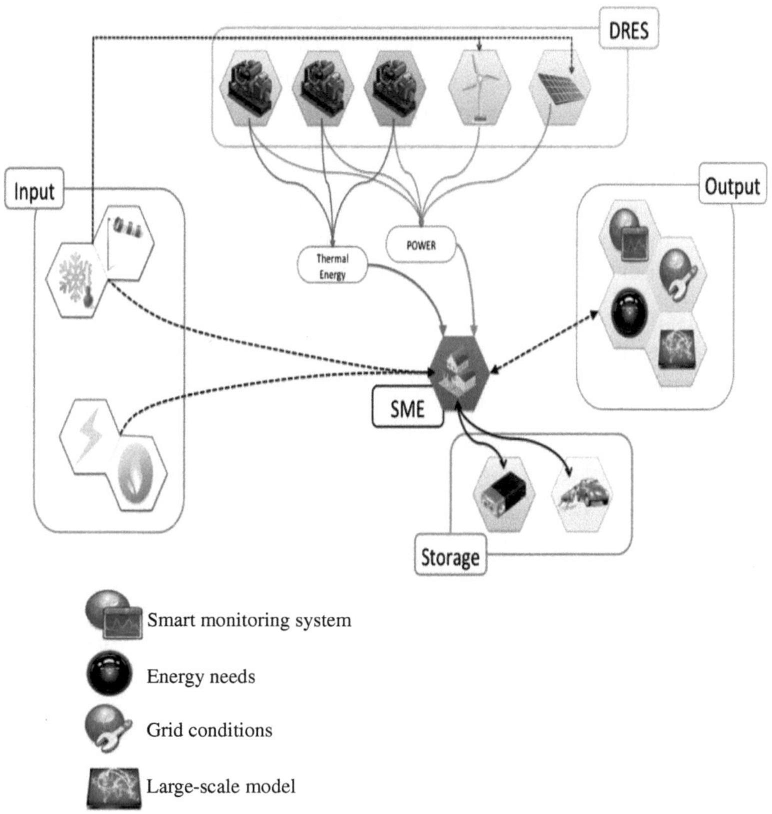

Fig. 10.1 The elements of the modeled Micro-Grid and their interconnections. *Source* Caló [1]

Smart grid-related empirical applications have also been developed. Caló [1] for instance has developed an interesting application of a hybrid Micro-Grid network. This grid makes smart use of the available renewable energy resources based on a demand-response logic principle. This hybrid system includes a number of selectable energy production units: 3 bioenergy-based combined heat and power (CHP) units, a wind turbine, and a photo voltage (PV) system. It also uses a plug-in hybrid electric vehicle (PHEV) as one of the storage elements. Figure 10.1 presents a schematic representation of the system components and their interconnections.

The main point of the simulator is to model the communication with the energy network from the end-user's point of view, serving as a blueprint for the modeling of a smart energy network. At this first stage of the development, communication with the power grid has been monitored, however, the modular structure of the simulator makes it easy to add further components. The main components of the system architecture are:

- "The SME block, containing the description of the built environment with the corresponding need of electric and thermal energy
- The Input block, providing the necessary environmental and system information.
- The Distributed renewable energy sources (DRES) block with the description of a number of selectable energy production systems, some of which related to the environmental data.
- The Storage block describing the electric energy storage device, and a sub-block considering the possibility to of a plug in hybrid electric vehicle.
- The Output block where all the necessary data are collected and monitored" [1].

The results from Caló model are based on varying the inclusion of different energy sources to the system. All the considered scenarios were based on predetermined common power and thermal energy requirements. All the reported data covered a period of one year, from the zero hour of the first of January to the midnight of the 31st of December 2010.

The simulations illustrate the potential of interplay among different energy vectors. An example of these aspects is, for example, the use of a PHEV as a possible household energy storage. From the power consumption point of view, it appeared as if the use of a PHEV brought no benefit to household energy budget. A closer look, on the other hand, revealed a substantial reduction in fuel consumption, providing an environmental and an economic benefit otherwise undetectable analyzing merely the overall power consumption [1].

10.4 Future Electricity Markets

All components of the electricity system—retail, generation, transmission, and distribution—are going to be under heavy stress for changes in the future. Diversity can be seen as the main driver of these changes. In generation, diversity is going to increase from potentially increasing competition. Increasing competition can result from more internationally oriented future power market and from more players in the local distributed generation market. Many lessons concerning deregulation and liberalization have been learned and it can be assumed that opening generation more to competition combined with an increase in the number of suppliers (at least on the local level) can result in considerable efficiency gains.

Also the role of renewable energy sources is going to increase strongly in the future. Forecasts for the share of wind power for instance are really massive. Although the cumulative capacity for wind power grew, there was a fall in annual additions as major wind markets, such as the US, Germany, and Spain faced economic problems following the global economic crisis. The global wind power markets are expected to recover due to the huge order intake by major wind manufacturers, the growing Asia–Pacific region, emerging South America and Africa regions, steady European wind markets, and recovery in North America.

The growth of major wind power markets (the US, Germany, Spain, France, Italy, India, and China) is, however, expected to slow down. Emerging markets from Asia–Pacific and South and Central America will gain a considerable market share. The growing Asia–Pacific wind power market powered by India, China, and other emerging countries such as Republic of Korea, Thailand, and Philippines will continue to drive the market in the region. Countries, such as Argentina, South Africa, Philippines, Ukraine, Brazil, Republic of Korea, and Mexico are some of the nascent wind markets which are set to expand rapidly in the forecast period.

Related to the diversity change distributed generation is going to change the scale of generation completely. Through intelligent networks small-scale generation can be included into the system. At the same time there is going to be a change from high-voltage toward low-voltage generation. Taking current developments into account and based on the results of the questionnaire-based survey (38 utilities organizations in 11 European countries) by Donkelaar and Scheepers [3] the three main features of future electricity distribution system can be summarized as follows. First, distribution networks of the future are likely to be managed actively with considerable amount of computer, communication, and control technologies applied to manage physical flows on the network as well as the flows of information between various devices controlling the behavior of the plant and equipment. Second, Distribution System Operators will have to take more responsibilities for the provision of security-related services. This would be a new task, which Distribution System Operators would need to conduct. This will be necessary if various forms of distributed generation are to be integrated in the operation and development of the entire system in order to ensure its secure operation and adequate service quality. An increasing penetration of distributed generation could potentially challenge the fundamental paradigm of central management of system security. With a very large penetration of small-scale generation (millions of various units), i.e., with the increased number of independent decision-making entities, a radical change from the central to a distributed management of the entire system operation will be required. Third, this technical challenge will, in turn, impose serious questions as to what market and commercial arrangements are needed to manage the balance between demand and supply in a system composed of millions of small generators and what regulatory approaches would facilitate evolution of the system from its present to its future form.[2]

Concerning transmission, reversal of all kinds of bottlenecks is one of the primary targets. Increase in international integration of energy systems necessitates that the transmission lines are capable of transmitting cross-border demand and supply. In 2010, transmission system operators and power exchanges from the Central Western European countries (CWE), Nordic countries, and European Market Coupling Company (EMCC) initiated the integration of the CWE and Nordic markets. EMCC was asked to develop joint business processes and to setup the market coupling system to integrate the two largest regional power markets. Interim Tight Volume Coupling is based on the previous Nordic-German tight

---

2 The original list of the summarized feature is presented in Donkelaar and Scheepers [3, pp. 31–32].

volume system operated by EMCC. At the same time the quality of transmission lines has to be secured in order to avoid cutbacks. This also makes it possible to open also transmission lines to contestability and possibly to competition.

The basic driver of the distribution and transmission-related changes is the technological change leading to intelligent networks. This development is going to open the current one-way traffic to an interactive dynamic two-way market place. Homes and other real estate are going to be monitored and optimized through computers and interconnected devices. The change is going to be very much like the one that has already happened in security services. In addition to being monitored for real life burglars our houses are going to be monitored for energy "thieves" which can be seen to be e.g., non-optimal heating, lights being on without anyone needing them, or washing machines running while the electricity price is high. At the same time our houses are connected to local cells and energy demand and supply is optimized in these cells through virtual operators. These changes necessitate the emergence of new business logic designed for these new markets. The basic driver in this business logic dimension is the move toward new types of services. There is going to emerge strong need for new types of services related to installing new devices and educating their use, drawing, and accepting new types of contracts relating to electricity prices and buying and selling of electricity. And of course, the new concept of virtual power plant must be complemented with all kinds of new services. Needless to say, these changes also put heavy stress on developing totally new types of intelligent incentive-based regulation models.

In the beginning of this book in Chap. 2 we discussed on six issues on which the success of restructuring and deregulation process, and the target to improve efficiency of electricity industry depends on. These six issues were, *number of active players* in the market, the *rules of the bidding procedure in the wholesale market*, the *organization of the demand side operation* in the market, the role of the *transmission grid as a neutral market place*, the mixture of *production technologies*, and *the ownership structure* of the utilities.

All of these issues are important also in the future and as a conclusion we discuss each of these six issues in this future perspective. In the future not only the *number of active players* but also the different roles of active players are increasingly important. It is likely that we see even bigger players (as a result of internationalization) in the market but it is also likely that the amount of small player is increasing as a result of distributed generation. Both roles should be seen important and necessary. The challenge is to organize this new market efficiently, in an environmentally sustainable manner and by assuring the system security.

In the future we continue to need efficient and well functioning *rules of the bidding procedure in the wholesale market.* Because of the increasing role of the distributed small-scale generation we, however, have a new challenge on deciding how these local market are connected to the international bidding procedure of the wholesale markets.

In the future the o*rganization of the demand side operation* creates various new possibilities which may have significant efficiency and welfare increasing effects. Economists have long discussed on the mechanisms which can improve demand side operations but not until now the required technologies have been available.

We have analyzed one of these new mechanisms, Real-Time Pricing, earlier in this book. The challenge of future demand side management is in getting all market participants to exploit these new mechanisms in their activities.

The *role of the transmission grid* as offering a neutral market place for competitive activities continues to be important also in the future. However, the role of the distribution network is going to increase as a result of increased role of the distributed generation. This creates new challenges for the regulator, system operator, and local market participants. Many of these challenges and the changing role of distribution network is already discussed above.

The role of the *mixture of production technologies* is going to be even more important in the future. In the future the energy production is going to be more distributed with various technologies. The concern of climate change and the continually increasing emissions require changing the energy production from fossil fuel-based production to the renewable energy sources-based production. Also the importance of energy saving technologies, energy efficiency in building and energy conservation by households is going to increase. It is likely that also the *ownership structure* of energy suppliers is going to be more diversified starting from single households and local communities to large international companies.

Finally we can say that worlds' energy markets are perhaps in the most challenging situation since the early stage of electrification. These challenges can be seen as threats but also they can be seen as a really new opportunity to create more sustainable, intelligent, and efficiently functioning Modern Energy Markets.

References

1. Caló A (2011) The future of energy services. Potential of smart energy networks in the northern periphery. Report on MicrE micro energy to rural enterprise. Thule Institute, Oulu, Finland
2. Ten Donkelaar M (2004) A survey of solutions and options for the integration of distributed generation into electricity supply systems. Energy Environ 15:323–332
3. Donkelaar M Ten, Scheepers MJJ (2004) A socio-economic analysis of technical solutions and practices for the integration of distributed generation. ECN Policy Studies, ECN—04-011. Available at http://www.ecn.nl/docs/library/report/2004/c04011.pdf
4. Holt LW, Kury TJ, Berg SV, Jamison MA (2009) The changing electric system architecture. Available at SSRN: http://ssrn.com/abstract=1534728
5. Knab S, Strunz K, Lehman H (2010) Smart grid: the central nervous system for power supply—new paradigms, new challenges, new services. Technische Universität Berlin, Innovationszentrum Energie, Schiftenreiche Nr. 2
6. Lasseter R, Akhil A, Marnay C, Stephens J, Dagle J, Guttromson R, Meliopoulos AS, Yinger R, Eto J (2002) Integration of distributed energy resources: the certs microgrid concept, April, LBLN 50829
7. Overbeeke F, van Roberts V (2002) Active networks as facilitators for embedded generation. Cogeneration and On-Site Power Production, Online Magazine 3(2)
8. Rochet J-C, Tirole J (2006) Two-sided markets: a progress report. Rand J Econ 37:645–667
9. Sullivan M, Schellenberg MA (2010) How to asses the economic consequences of smart grid reliability investments. Freeman, Sullivan & Co, USA

10. Sullivan M, Schellenberg J (2011) Smart grid economics: the cost-benefit analysis. renew grid magazine, April. Zackin Publications Inc
11. U.S. Department of Energy (2008) The smart grid: an introduction. Department of Energy, U.S, Washington
12. U.S. Department of Energy (2008) Smart grid system report. Department of Energy, U.S, Washington

Index

M. Kopsakangas-Savolainen and R. Svento, *Modern Energy Markets*, Green Energy and Technology, DOI: 10.1007/978-1-4471-2972-1,